GENETIC IMPROVEMENT OF CROPS

Emergent Techniques

GENETIC IMPROVEMENT OF CROPS

Emergent Techniques

Irwin Rubenstein
Burle Gengenbach
Ronald L. Phillips
and
C. Edward Green
Editors

University of Minnesota Press □ Minneapolis

Published by the University of Minnesota Press,
2037 University Avenue Southeast, Minneapolis MN 55414
Printed in the United States of America.

Library of Congress Cataloging in Publication Data

Main entry under title:

Genetic improvement of crops.

"Papers from a symposium held in 1979 at the University of Minnesota."

Includes index.

1. Plant-breeding—Congresses. 2. Plant genetics—Congresses. I. Rubenstein, Irwin. II. Minnesota. University.

SB123.G39 631.5'3 80-23560
ISBN 0-8166-0966-7

Preface

Over the last few years, a number of in vitro techniques have emerged that hold promise for the genetic alteration of higher organisms by nontraditional means. These new techniques include recombinant DNA and cloning, genome characterization, gene and organelle transfer, and plant tissue culture. These methods eventually should allow the study and possible modification of the genetic structure of economic crop plants. Because of the increasing demand for improved quantity and quality of plant products, it is imperative that all available avenues for crop improvement be used. Plant breeding, genetics, and cytogenetics will continue to supply major impetus to crop improvement advances, but the emergent in vitro technology also needs to be brought into the mainstream of plant improvement research. This task will be difficult and will require decades of effort. Its accomplishment will require numerous attempts to find potentially successful applications, as well as a vast increase in our basic knowledge of plant biology.

This volume contains papers from a symposium, held in 1979 at the University of Minnesota, that discussed emergent techniques for the genetic improvement of crops. This was the sixth symposium at Minnesota that focused on new trends in biology of particular importance to plant improvement. The meeting was attended by approximately 200 people from various universities and industries.

The symposium started with a session on the potentials for crop improvement and the genetic needs of plant breeders. This session set the tone for the symposium. The following sessions covered topics in genome organization and function, recombinant DNA technology and application, gene transfer and expression, organelle transfer, and plant tissue culture.

The editors gratefully acknowledge financial support from the National Science Foundation; at the University of Minnesota, the College of Biological Sciences, the Department of Genetics and Cell Biology, the Department of Agronomy and Plant Genetics; the NIH Training Grant: Cellular and Molecular Biology; the University of Minnesota chapter of Sigma Xi (the scientific research society of North America); Pillsbury Company; and the Red River Valley Potato Growers' Association.

Contributors

Athwal, R. S., Department of Microbiology, New Jersey School of Medicine, 100 Bergen Street, Newark, N.J. 07103.

Bedbrook, J., Department of Cytogenetics, Plant Breeding Institute, Trumpington, Cambridge, England CB2 2LQ. Current address: CSIRO, Division of Plant Industry, P.O. Box 1600, Canberra City, Act 2601, Australia.

Belford, H. S., Plant Biology Department, Carnegie Institution, Panama Street, Stanford, Calif. 94305.

Bonnett, H., Biology Department, University of Oregon, Eugene, Oreg. 97403.

Chilton, M.-D., Department of Microbiology, University of Washington, Seattle, Wash. 98195. Current address: Department of Biology, Washington University, Box 1137, St. Louis, Mo. 63130.

Cuellar, R. E., Plant Biology Department, Carnegie Institution, Panama Street, Stanford, Calif. 94305.

Duvick, D. N., Pioneer Hi-Bred International, Johnston, Iowa 50131.

Flavell, R. B., Department of Cytogenetics, Plant Breeding Institute, Trumpington, Cambridge, England, CB2 2LQ.

Garfinkel, D. J., Department of Microbiology, University of Washington, Seattle, Wash. 98195.

Gelvin, S. B., Department of Microbiology, University of Washington, Seattle, Wash. 98195.

Gerlach, W., Department of Cytogenetics, Plant Breeding Institute, Trumpington, Cambridge, England, CB2 2LQ.

Glimelius, K., Biology Department, University of Oregon, Eugene, Oreg. 97403.

Gordon, M. P., Department of Microbiology, University of Washington, Seattle, Wash. 98195.

Jones, J., Department of Cytogenetics, Plant Breeding Institute, Trumpington, Cambridge, England, CB2 2LQ.

Kleese, R. A., Pioneer Hi-Bred International, Johnston, Iowa 50131.

Lutz, J., School of Life Sciences, University of Illinois, Urbana, Ill. 61801.

McBride, O. W., Department of Microbiology, New Jersey School of Medicine, 100 Bergen Street, Newark, N.J. 07103.

McPherson, J., Department of Microbiology, University of Washington, Seattle, Wash. 98195.

Meins, F., Jr., School of Life Sciences, University of Illinois, Urbana, Ill. 61801. Current address: Friedrich Meischer Institut, P.O. Box 273, CH-4002, Basel, Switzerland.

Merlo, D. J., Department of Microbiology, University of Washington, Seattle, Wash. 98195.

Montoya, A. L., Department of Microbiology, University of Washington, Seattle, Wash. 98195.

Murray, M. G. Plant Biology Department, Carnegie Institution, Panama Street, Stanford, Calif. 94305.

Nester, E. W., Department of Microbiology, University of Washington, Seattle, Wash. 98195.

Nutter, R. C., Department of Microbiology, University of Washington, Seattle, Wash. 98195.

O'Dell, M., Department of Cytogenetics, Plant Breeding Institute, Trumpington, Cambridge, England, CB2 2LQ.

Saiki, R. K., Department of Microbiology, University of Washington, Seattle, Wash. 98195.

Shepard, J. F., Plant Pathology Department, Kansas State University, Manhattan, Kans. 66506.

Strobel, G. A., Plant Pathology Department, Montana State University, Bozeman, Mont. 59715.

Thomashow, M. F., Department of Microbiology, University of Washington, Seattle, Wash. 98195.

Thompson, R., Department of Cytogenetics, Plant Breeding Institute, Trumpington, Cambridge, England, CB2 2LQ.

Thompson, W. F., Plant Biology Department, Carnegie Institution, Panama Street, Stanford, Calif. 94305.

Wallin, A., Biology Department, University of Oregon, Eugene, Oreg. 97403.

Yang, F.-M., Department of Microbiology, University of Washington, Seattle, Wash. 98195.

Contents

Part I: Potentials and Needs

Potentials for Improving Crop Production

G. A. Strobel

The needs of humans for food, fiber, fuel, and medicine are outstripping their supply. New ideas need to be explored in all areas of crop production—from plant introduction to genetic manipulation and molecular biology. Gains in productivity of many major food plants are becoming harder to obtain. We cannot expect to continue to make major strides in improving crop production by employing strictly conventional approaches and techniques.

Limitations and Concerns in Crop Productivity

To establish reasonable goals for crop improvement, we must know the limitations of production. In many of our major crops, productivity could be increased greatly if pests (diseases, insects, weeds) could be controlled. The magnitude of loss is not known precisely for all crops, but Coramer (1967) and the Food and Agriculture Organization (1975) have conservatively estimated preharvest losses

Note: The work reported herein was supported in part by a grant from the Herman Frasch Foundation, NSF grant PLM 7822517, and the Montana Agricultural Experiment Station.

for some food crops (Table 1). Losses after harvest caused by vertebrates such as birds, rats, and other rodents also take a toll. There are three major ways that pest control can improve crops: (1) by increasing yield, (2) by decreasing destruction in storage, and (3) by allowing options of growing a particular crop where it could not otherwise be grown. Pest problems involve the interaction of at least two biological systems, and genetic variation is inevitable. Effective control can result only from constant monitoring of crops and their pest populations. The genetic approach is the most economical and least damaging to the environment compared to establishing cultural practices, integrating pest-management strategies, utilizing pesticides, or developing biological control methods.

The development of plants with increased tolerance of adverse environmental conditions offers considerable promise for crop improvement (National Research Council, 1977). Specific stress factors include soil aluminum, heat, cold, and drought. Drought has been an annual problem throughout history. The most recent devastating drought occurred during the mid-1970s in Europe, North America, Africa, and Asia. Crop improvements could be made by reducing the risks presented by drought by developing faster maturing varieties and varieties with more extensive root systems and selecting for varieties whose leaf anatomy is better adapted to dry conditions. Furthermore, the crop's ability to resume rapid growth following drought stress may be vital in determining final yield.

Table 1. Losses from Pests in the World's Major Crops

Crop	Pest Losses %	1974 Production in Millions of Tons	Potential Production
Rice (paddy)	46.4	323	603
Wheat	23.9	360	473
Maize	34.8	293	449
Millet	38.0	93	150
Potatoes	32.3	294	434
Cassava	33.5	104	158
Sweet potatoes	25.5	134	180
Tomatoes	24.5	36	48
Soybeans	29.1	57	80

The 3% average annual increase in grain production during the past 25 years was dependent largely upon the exponential use of nitrogen fertilizer (National Research Council, 1977). Natural gas is the preferred source of reducing power in the production of ammonia nitrogen. By the end of this century, at current levels of ammonia consumption, it is estimated that at least 10% of the natural gas supply will be utilized for making fertilizer.

All plants require nitrogen to survive. Legumes and a few other plants and microorganisms fix atmospheric nitrogen and satisfy their own needs. Surpisingly, legume yields could be increased substantially if nitrogen fixation rates in the nodules could be increased. If cereal crops could be engineered to fix nitrogen at 30% of the soybean rate (i.e., 25 kg N_2/ha), the effect on yield would be equivalent to the total increase in yield obtained by the worldwide use of fertilizer in 1975 (National Research Council, 1977). Enhanced biological nitrogen fixation would conserve fossil fuels, increase yields, and save the consumer money.

One wasteful process that undoubtedly limits plant productivity is photorespiration. Typically, plants lacking this process are the fastest growing and highest yielding food crops; sugarcane and corn are examples. Ideally, selection or manipulation of crop plants such as rice, wheat, and barley to eliminate or reduce the photorespiratory process would be desirable. Of course, other physiological and biochemical parameters such as solute transport, control of conversion of vegetative to reproductive growth, and premature senescence may limit productivity.

Overriding all of the physical and biological limitations of foodcrop improvement is the nutritive quality of the final product. Genetic manipulation coupled with appropriate selection techniques is a useful approach.

New Crops

Of the approximately 300,000 known plant species, fewer than 300 are used in organized agriculture. In the United States, 100 species are grown commercially at economic levels above $1 million (Princen, 1977). With proper stimulation, exploration, and research, other wild plants probably could be domesticated. Most major crops of the United States are introductions, with the sunflower as a notable exception. Plant introductions into this country have been made since the seventeenth century. In one instance, a ship from Madagascar made an unscheduled docking in South Carolina in 1694. The ship's captain presented the colonists with a handful of rice. Today, rice is an important export crop of the United States.

Overall, 10 crops (rice, wheat, corn, barley, millet, sweet potatoes, potatoes, soybeans, oats, and cassava) constitute the basic food resources for the world's peoples. These crops often are grown as a monoculture that makes them extremely vulnerable to pests. Notable examples are the late blight epidemic of potatoes in Ireland in the mid-1840s and the Southern corn leaf blight epidemic in the United States in 1970. More crop diversity in the world's food-producing countries is needed.

New developments in tropical agriculture have not occurred largely because the major centers of scientific research are located in temperate zones. A staggering variety of plant species could supply a wealth of products, such as oils, gums, proteins, drugs, and medicinals. Innovations in transport, consumer affluence, industrial processes, and consumer demand are good reasons for exploiting tropical plant (National Academy of Sciences, 1975). Given a concentrated research effort, many tropical plants could follow the course of the soybean, which was an oddity in the United States 50 years ago. Some specific tropical plants with agronomic potential are described below.

Mangosteen

The fruit of this plant, often described as the world's best flavored, is highly esteemed in Southeast Asia but is not available in potentially major market areas in Europe and North America. It could be produced commercially in areas of the world having the proper growing period, temperatures between 20°C and 30°C, and deep, fertile, well-drained soils. Horticultural problems, such as shortening the time necessary to produce mature plants and developing vegetative propagation methods remain unsolved.

Naranjilla

The native homes of this solanaceous plant are Colombia and Ecuador. The fruit makes an excellent dessert fruit and can also be used to flavor confections. The fruit is spherical, yellow-orange, and sometimes as large as a tennis ball and it has a yellow-green pulp containing numerous seeds. The plant produces fruit year-round. With new information on pest-control practices combined with varieties that would widen its climatic distribution, naranjilla could be grown throughout the tropics.

Pejibaye or Peach Palm

This member of the Palmae could be an important commercial crop throughout the humid tropics. It appears to be the most balanced of all tropical foods since its fruit contains high levels of carbohydrates, proteins, lipids, minerals, and vitamins. It has twice the protein content of bananas and can provide a greater yield of carbohydrate and protein than corn. Because of its multiple stems, it is one of the most promising palms for the plantation production of hearts of palm. Basic horticultural work on varietal improvement and diffusion of seed and information on this crop could lead to its productions in many tropical areas.

Uvilla

This small tree of the western Amazon produces racemes of purple, grapelike fruit with large pits nad sweet, white pulp. It grows prolifically and fruits heavily over a 3-month period. Uvilla has never been the subject of scientific study.

Other Plants

Numerous other tropical plants have agronomic potential including pummelo, durian, guinua, cocoyams, wax gourd, and chaya (National Academy of Sciences, 1975).

Producing New Tropical Food Crops

Many of the developing countries, where food is desperately needed, lie within the tropics. In the past, the people of these nations have been fed on the surplus food grains of North America and elsewhere. Research on the technology of tropical agriculture has been thwarted by a lack of money and a false sense of security promoted by the giveaway programs of the producing countries. Because of the increased costs of production and the lowered prices in the marketplaces, many of the producing countries are cutting production. Produce is available to the poor countries, but for a price. If they could develop their own agriculture, their products would be in demand elsewhere. Some think that the advantage for food production lies in the tropics, where sunlight, moisture, and a sizable, inexpensive work force are plentiful.

Temperature and subtropical climates still hold tremendous potential for the introduction of new crops. Some potential crops include kenaf (a fast-growing plant used for pulp), crambe (a seed used for oil and protein), vernonia (a seed used or oil), stokesia (a seed used for oil), and jojoba (used for oils, waxes, alcohols, and acid derivatives).

Petroleum Plantations

Currently, we use 98% of our energy in the form of fossilized carbon. Calvin (1978) has suggested that certain plants may be able to supply humans with compounds for energy purposes. The advantages of using a renewable energy source are: no net increase in our atmospheric CO_2 level and a decrease in the level of carcinogens spewed into the environment. Within 5 years, Brazil plans to be producing 20 billion liters of alcohol per year for auto fuel and a source of chemicals for the petrochemical industry. The plant supplying the carbon skeletons for this task is sugarcane. In colder climates, Calvin (1978) proposes that we develop plantations with latex-bearing species of Euphorbia, such as *Euphorbia tirucalli, E. lathyris* (Fig. 1), and *E. lactea*. In field trials in California, the fast-growing *E. lathyris* yielded 10 barrels of oil per acre in 9 months (Fig. 2). This oil is a mixture of hydrocarbons, mostly

Fig. 1. *Euphorbia lathyris* growing in a test plantation in California. (Courtesy of Lawrence Radiation Lab, Berkeley, California.)

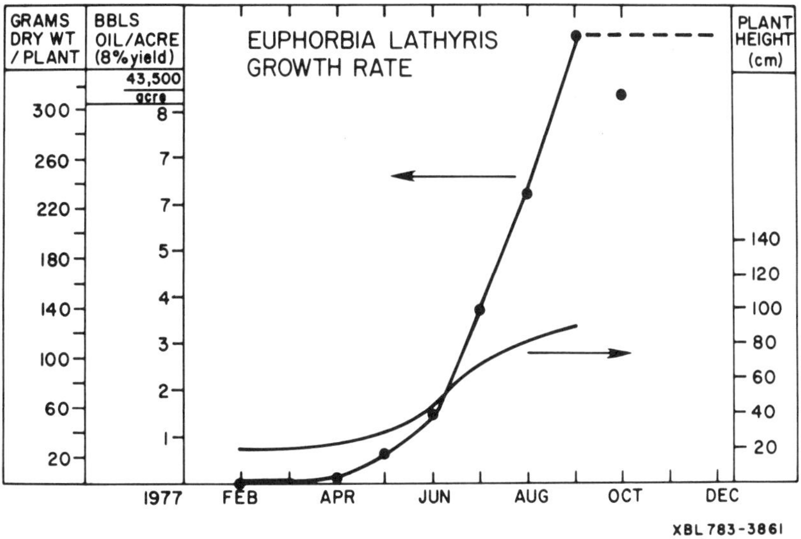

Fig. 2. The growth rate of *Euphorbia lathyris* and its yield of oil/acre. (Courtesy of Lawrence Radiation Lab, Berkeley, California.)

terpenoid isoprenes, that is, open-chain and cyclic isoprenes, diterpenes, triterpenes, sterols, et cetera. The total cost of growing the plant and extracting and canning the oil is estimated to be $20 per barrel. Selection, genetic improvement, and domestication of these Euphorbs would increase the yield of oil substantially. The oil could easily be processed to yield gasoline and products for the petrochemical industry.

Plants as a Source of Drugs

Humans first discovered and developed the cereals, then the legumes, and now the plants that yield drugs. At least 25% of the prescriptions dispensed in the United States contain one or more active ingredients extracted from plants (Farnsworth, 1977). Compounds in plants have been used to treat virtually every illness known (Lewis and Lewis, 1977). Recently, the list of biologically active compounds isolated from plants has grown considerably.

Plant Breeding—Unconventional Strategies

The tremendous increase in corn production in the United States over the past 40 years is, in part, directly related to the development of hybrid corn. Unlike corn, many other cereals are not openly cross-pollinated. Therefore, hybrids are more difficult to obtain in them. Nevertheless, male sterility in these cereals is potentially important in varietal improvement. A male-sterile-facilitated recurrent-selection program at Montana State University illustrates this point. The ultimate goal is to "pyramid" genes for fine agronomic qualities and disease resistance in barley. Eventually, these barley cultivars will be released in the developing countries. In this method for barley improvement, 6 to 8 male sterile lines with excellent agronomic qualities are interplanted with 15 or more diverse but widely adapted cultivars with broadly based disease resistance (Fig. 3). The F_1's from this cross are selfed and yield about 25% male-sterile plants. These can be crossed with other disease-resistant cultivars to accumulate genes (pyramiding) for

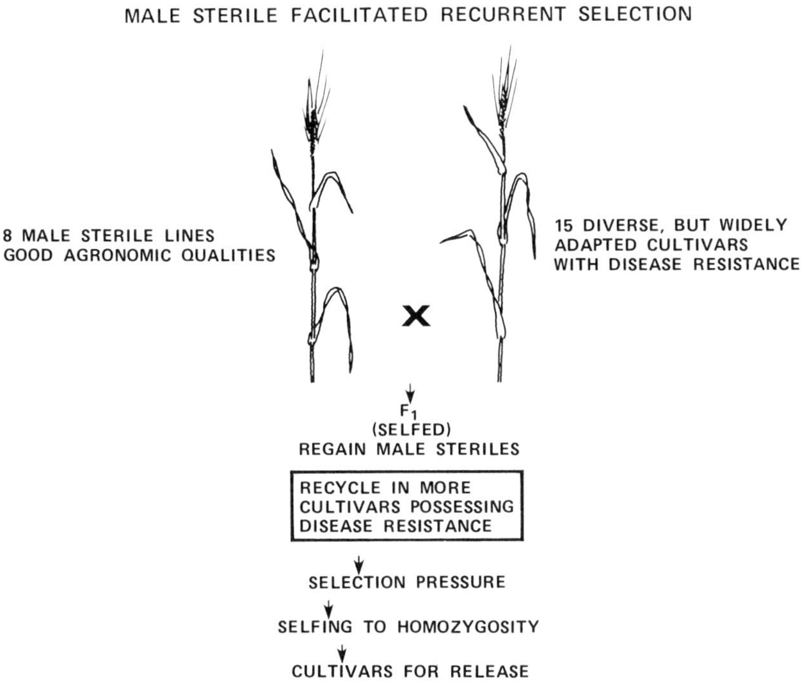

Fig. 3. Diagram of the male sterile facilitated recurrent-selection technique for pyramiding genes for disease resistance in barley.

disease resistance. Adequate selection tests are required throughout this program. Eventually, desirable plants can be selfed and used as agronomic cultivars or simply as a gene source in other breeding programs. A recurrent-selection program for crop improvement has enormous potential for other important food plants.

Bursts of yield improvement for oats (Langer et al., 1978) have been made only when there has been an introgression of germplasm from other species or ecotypes. Thus, the continued discovery, preservation, and use of germplasm from other species, ecotypes, and variants of important food crops that may serve as new sources of genes governing yield, disease resistance, and drought tolerance is vital. The discoveries of the "Gigas" form of bread wheat (Atsmon and Jacobs, 1977) and a new wheat species named *Triticum searsii* by M. Feldman of the Weizmann Institute are two recent examples. The "Gigas" type has a spike and vegetative parts impressively larger than standard cultivars (Fig. 4). The new wheat species *T. searsii* possesses only the B genome, whereas all commercially grown bread wheats possess the AB and D genomes. The B genome is significant to wheat breeders because it contains a deleterious gene that precludes the pairing of homoeologous chromosomes. Thus, the antipairing trait has hindered progress in wheat improvement since genetic mixing has not been maximized. If individuals in the *T. searsii* population could be found that lack or are modified in the antipairing trait, new bread wheats could be produced that might have higher yields, better disease resistance, and drought tolerance.

Recently, Iltis (Sullivan, 1979) has found a new maize species in Mexico (named *Zea diploperennis*) that has 20 chromosomes and appears to produce fertile plants when crossed with corn, *Zea mays*. The new maize has 20 chromosomes, is a perennial, and, thus, might yield hybrid corns that produce fruit year-round. The native home of *Zea diploperennis* is a small, mountainous region of Mexico. There is a definite need to set aside and preserve the habitat of potentially important germplasms.

The possibility of practical results from intergeneric hybridization of cereals is best exemplified by triticale, a wheat-rye hybrid. Sebesta (personal communication) has prepared triticale types that carry green-bug-resistance genes from the rye parent. Pollen from the hybrid is subjected to x-radiation and then backcrossed to wheat. One line, Amigo, that has been released carries one rye-chromosome translocation in an otherwise complete wheat genetic background. The potential of transferring disease resistance and other desirable features from rye into triticale and eventually into wheat is enormous.

Barley-wheat hybrids could conceivably possess the salt and drought tolerance of barley and the protein and yield qualities of wheat. Successful wheat-barley hybrids apparently have been made. Mujeeb and his associates (1978) have crossed *Hordeum vulgare* ($2n = 14$) by *T. turgidum* ($2n = 4x = 28$) and *T. aestivum* ($2n = 6x = 42$). Predominantly, the hybrid plants have 21 chromo-

Fig. 4. A "Gigas" spikelet on the right compared to a normal wheat spikelet on the left.

somes, all contributed by the female barley parent. Thus, paternal-chromosome elimination integrated with somatic doubling of the maternal-chromosome complement apparently has occurred. Nevertheless, the "hybrids" have some characteristics of the paternal wheat parent. Mujeeb and his associates (1978) refer to this phenomenon as "microhybridity." Biochemical, pathological, and molecular biological tests will allow them to determine whether DNA transfer from the parental wheat had actually occurred versus "switching on" of previously suppressed barley DNA. Bates and his associates (1976) also consider the chemical pretreatment of the maternal parent for hybridization in a wide cross, and it may be that getting microhybridity for an advantageous gene ultimately holds some potential for crop improvement.

Vegetative Cloning Techniques

The grafting of superior scion stock to disease-resistant root stocks and the acquisition of higher yielding virus-free plants by meristem-culture techniques have been extremely useful procedures in plant improvement. However, plant-callus techniques and cell- and tissue-cloning techniques have not yielded practical benefits for crop improvement until recently. Among crop plants, tobacco, asparagus, rape, and potato have been regenerated from protoplasts (Gamborg, et al., 1978). In general, we have learned from protoplast work that each species seems to be unique in its regeneration requirements and that protoplasts from various species can be fused and that they undergo various amounts of cell division. The more closely related the species, the more likely it is that plant regeneration from the hybrid fusion product will result (Carlson et al., 1972). Fusion of protoplasts, facilitated with polyethylene glycol to produce genuine hybrids, may improve crops in a manner analogous to the wide crosses made with cereals (for example, triticale). Gamborg and Holl (1977), at the Prairie Regional Laboratory in Canada, have made a number of fusion products from different species. However, a major problem has been the lack of morphogenesis of the fused products, especially those made between widely diverse species (Fig. 5).

The successful regeneration of potato plants (Russet Burbank) from leaf-mesophyll protoplasts offers a major opportunity for the improvement of this crop (Shepard and Totten, 1977). Because Russet Burbank is a tetraploid, improvement in pest resistance and other qualities of this important potato variety by traditional plant-breeding techniques has been virtually impossible. In their initial experiments, Shepard and Totten (1977) showed that nearly 25% of the regenerated plants were phenotypically different from their mother plants. The results could not be accounted for by external mutagens. Matern, Strobel, and Shepard (1978) demonstrated that the regenerated clones reacted differentially to *Alternaria solani*, the causal agent of early blight. Some clones were susceptible, some were intermediate, and others were resistant. The mother clone was susceptible. These clones behaved in a man-

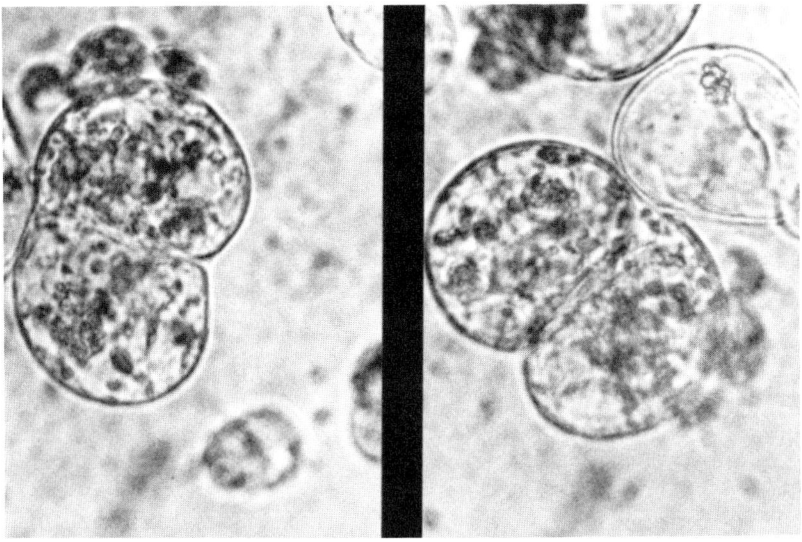

Fig. 5. Protoplasts of corn and sorghum that have fused and are undergoing cell division. (Courtesy of D. Brar.)

ner comparable to toxin preparations obtained from the fungus (Fig. 6). Ideally, these toxins could serve as a useful tool in screening protoplasts for disease resistance. The genetic nature of this wide phenotypic variation is not understood.

The most immediate practical benefits from protoplast-regeneration techniques are likely to occur with the vegetatively propagated crop plants, such as potato, sweet potato, sugarcane, and strawberry. Variation, either of the type observed in potato or induced by applied mutagens, may yield one or more desirable plants that could then serve as the stock for commercial propagation. However, an inherent problem in any tissue-, embryo-, or protoplast-culture method is the availability of appropriate screening techniques. In addition to phytotoxins (Gengenbach et al., 1977; Carlson, 1973; Strobel, 1974; and Strobel, 1977), pesticides, environmental factors, and specific enzyme inhibitors are possible selection tools.

Fig. 6. The reaction of several potato leaves obtained from cloned potato protoplasts to the toxins of *A. solani.*

Organelle Transfer

Chloroplasts may have evolved from primitive, blue-green algae that were intracellular symbionts. Chloroplasts resemble photosynthetic prokaryotes genetically and cytologically. The small, circular genome of chloroplasts should be amenable to manipulation by the techniques of bacterial genetics. Since chloroplasts replicate semiautonomously (Kirk, 1971), alterations in the genome are heritable by daughter cells and progeny plants. Modifications of the photosynthetic apparatus may be engineered in the chloroplast genome since some of the protein subunits of carboxydismutase are coded by the chloroplast DNA. Engineering of these metabolic capabilities may be possible as in the following discussion simplified from that of Miller (1978):

1. Isolate and maintain active chloroplasts.
2. Introduce a foreign chloroplast DNA or a recombinant chloroplast DNA prepared by classical procedures.
3. Introduce, via fusion or micromanipulation, the chloroplast preparation into cultured plant cells (Bonnett, 1976). Cells

containing recombinant chloroplasts could be selected by toxin or drug resistance.

4. Regenerate the whole plant by standard procedures.

The chloroplast genome carries, in part, the information for RuBP carboxylase. Besides catalyzing the CO_2 fixation reaction, it also possesses oxygenase activity leading to a photorespiratory loss of CO_2. By abolishing photorespiration, photosynthesis could be enhanced in excess of 40% (Laing, et al., 1974). Therefore, it seems that a potential benefit to plant productivity could come via the transfer of chloroplasts that carry information for a more efficient RuBP carboxylase. Since chloroplast uptake is a fairly efficient process (Bonnett, 1976), selection of the most efficient photosynthesis types could be done at the whole plant level (after regeneration). However, a test system is needed to determine adequately the relative efficiency of the CO_2 reaction. Fortunately, a new labeling scheme and the appropriate instrumentation for the quick determination of these two activities have been invented by Kent and his associates (1979). Ribulose diphosphate-[^{3}H]-C-5 and ribulose diphosphate-[^{14}C]-C-1 are administered to plant material. Oxygenase activity favors the production of only 1 mole of 3-phosphoglyceric acid plus 1 mole of phosphoglycolate, whereas carboxylase activity produces 2 moles of 3-phosphoglyceric acid (Fig. 7). The relative abundance of [^{14}C]:[^{3}H] in the 3-phosphoglyceric acid produced in the reaction is determined. This method should facilitate the discovery of more efficient carboxylases in nature or those produced by genetic-engineering technology.

The importance of extranuclear DNA in the control of male sterility in plants has been known for years. The vital importance of this phenomenon was not fully realized until the Southern corn blight epidemic in the United States in 1970, the largest plant-disease epidemic occurring in the shortest period in history. Prior to the epidemic, hybrids were made using cytoplasmic male sterility to effect the cross. Unfortunately, susceptibility to the causal fungus, *Helminthosporium maydis*, was also a cytoplasmic trait. Male sterility and sensitivity to the fungal toxin may not be inherited independently in the cytoplasm (Gengenbach et al., 1977). Since restriction-endonuclease maps of DNA isolate from

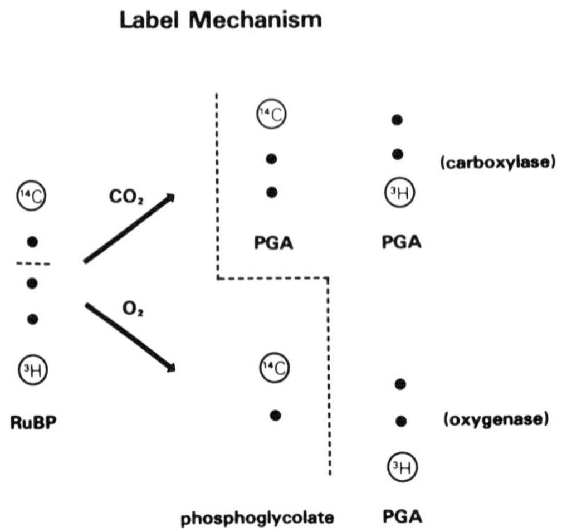

Fig. 7. The principle of the labeling test scheme for determining the relative carboxylase versus oxygenase activities of RuBP carboxylase. (Courtesy of S. S. Kent.)

mitochondria bearing N cytoplasm (resistant, fertile) differ markedly from the mitochondrial DNA from Tms cytoplasm (susceptible, sterile), these two characters could be controlled by mitochondrial DNA (Levings and Pring, 1976). The possibility could not be ruled out that chloroplast DNA could control these traits. Furthermore, other cytoplasmic DNAs similar to the plasmid DNA in bacteria, which have been implicated but not yet found in higher plants, may also be involved (McClintock, 1967). Such DNAs have been described as "a phenomenon without an investigator." Possibly, cytoplasmic traits carried on mitochondrial DNA or on other extrachromosomal DNA could be transferred to protoplasts and eventually be expressed in the regenerated plant. The nature of the information in the plant mitochondrial DNA must be more fully investigated as well as the function of the remaining 90% of chloroplast DNA (Ciferri, 1978). Furthermore, plant enzymes that may act as restriction endonucleases need to be searched out. It seems likely that they could play a key role in the modification of all DNAs in the cell.

Gene Transfer

A major obstacle to the incorporation of isolated DNA into plant materials (tissues, seeds, protoplasts) is the presence of enzymes that systematically degrade it. Many proposals to protect the donor DNA invoke the use of a gene carrier and successful methods from bacterial genetics. At least two types of DNAs, bacterial plasmids and plant viruses, can be considered as carriers (Hull, 1978). However, if other extrachromosomal DNAs of the type known in Saccharomyces (Guerrinean et al., 1974) are found in higher plants, the process of gene transfer might be facilitated by a third means.

Labeled DNA can be taken up by plant protoplasts (Okyama et al., 1972). However, integration of DNA into the host genome and the expression of new inheritable characteristics have not been confirmed.

Transformation of host cells via a bacterial plasmid occurs during gall formation in plants infected by *Agrobacterium tumefaciens* (Van Larebeke et al., 1974). Only a portion of the plasmid DNA is apparently integrated into the plant genome (Drummond et al., 1977). Schell and Montagu (1977) have proposed to engineer the transposition segments of DNA with important genes and DNA sequences and ultimately get them incorporated into the plant genome. Thus, the Ti plasmid is a potential tool whereby plants may be engineered by the modern technology of molecular genetics (see the discussion in *Science 203*, 254-55).

The Ti plasmid does not benefit the plant, so new desirable genes such as *Nif* genes must be incorporated into the Ti plasmid. Initially, the complexities of acquiring and then expressing plasmid recombinants are enormous. A more simplified system would be desirable. I propose that we consider the new plant-transformation system developed at Montana State University by Moore and his associates (1979). A large-molecular-weight plasmid (1.1×10^8 daltons) is responsible for the infectivity of this organism since (1) infectivity is lost when all or part of the plasmid is lost after treatment with ethidium bromide or heating at 37°C, (2) conjugational transfer to agrobacteria-lacking plasmids renders them infectious, and (3) restriction-endonuclease patterns of the plasmids of the transconjugants are identical to the wild-type donor bacterium.

Induction of rooting may be a practical use for the plasmid of *A. rhizogenes* (Fig. 8), designated pHr (A_4). The degree of root proliferation and drought tolerance are correlated in many species. For instance, Moore and Millikan (unpublished) showed that apple seedlings having the "hairy root" syndrome survived the 1975-76 drought in Missouri much better than their control counterparts, 93% versus 30%, respectively. Because the Hr plasmid causes root proliferation in its respective host, either it or an Hr-plasmid fragment may be a useful practical tool in promoting drought tolerance in certain plant species.

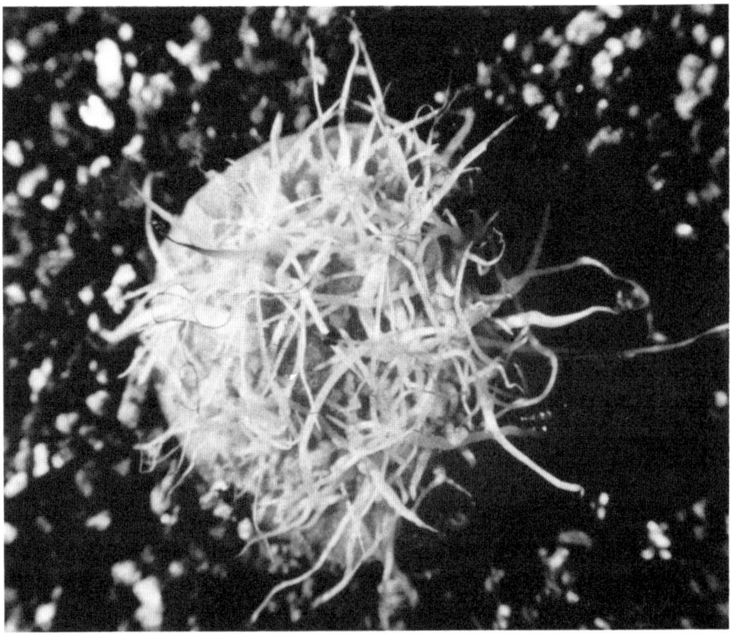

Fig. 8. Roots produced on a carrot disc 2 weeks after inoculation with 10^7 cells of *Agrobacterium rhizogenes.*

Summary

Creative ideas for improving crop production are needed to meet the demands of the world's population. The introduction and exploitation of new plants in tropical agriculture may have tremendous impact on the world's food, fiber, and energy production. The discovery of new germplasm in the epicenters of the world's major crops still may result in improved yields and pest resistance. Classical plant-breeding procedures with innovative approaches undoubtedly will be the mainstay of crop improvement for the next 10 years. Nevertheless, protoplast-culturing methods applied to vegetatively propagated plants such as potato and sweet potato have excellent potential for improving these crops. While transplantation of chloroplasts has been accomplished, more information is needed on the nature of the DNA in extrachromosomal organelles to make such experiments show potential for plant improvement. Gene transfer via bacterial plasmids or viral DNA is under study. Plasmid DNA of *A. rhizogenes* appears to transform plants by causing a proliferation of roots, which may confer drought tolerance. This system might be an ideal one for studying gene transfer.

References

Atsmon, D., and Jacobs, E. (1977) A newly bred "Gigas" form of bread wheat: morphological features and thermo-photoperiodic responses. *Crop Sci. 17*, 31-35.

Bates, L. S.; Mujeeb, K. A.; and Waters, R. F. (1976) Wheat-barley hybrids: Problems and potentials. *Cereal Res. Commun. 4*, 377-86.

Bonnett, A. T. (1976) On the mechanism of the uptake of Vaucheria chloroplasts of carrot protoplasts treated with polyethylene glycol. *Planta 131*, 229-33.

Calvin, M. (1978) *Petroleum Plantations.* United States Department of Energy, Berkeley, California. (Publication LBL 8236.)

Carlson, P. S. (1973) Methionine sulfoximine-resistant mutants of tobacco. *Science 180*, 1366-68.

Carlson, P.; Smith, H. H.; and Dearing, R. D. (1972) Parasexual interspecific plant hybridization. *Proc. Nat. Acad. Sci., U. S. A. 69*, 2292-94.

Ciferri, O. (1978) The chloroplast DNA mystery. *Trends in Biochem. Res. 3*, 256-58.

Coramer, H. H. (1967) *Plant protection and World Crop Production.* Farbenfabriken Bajer, Leverkussen, West Germany.

Drummond, M. H.; Gordon, M. P.; Nester, E. W.; and Chilton, M. D. (1977) Foreign DNA of bacterial plasmid origin is transcribed in crown gall tumors. *Nature 269*, 535-36.

Farnsworth, N. R. (1977) The Current Importance of Plants as a Source of Drugs. In *Crop Resources* (D. S. Seigler, ed.), pp. 61-73. Academic Press, New York.

Food and Agriculture Organization of the United Nations. (1975) *Production Yearbook, 1974.* Rome: Food and Agriculture Organization of the United Nations.

Gamborg, O., and Hall, F. B. (1977) *Plant Protoplasts Fusion and Hybridization in Genetic Engineering for Nitrogen Fixation* (A. Hollaender, ed.), pp. 299-316. Plenum, New York.

Gamborg, O. L.; Kartha, K. K.; Okyama, K.; and Fowke, L. (1978) *Protoplasts and Tissue Culture Methods in Crop Improvement.* Proceedings of a Symposium on Plant Tissue Culture, Peking, China, pp. 265-78. Science Press, Princeton, New Jersey.

Gengenbach, B. G.; Green, C. E.; and Donovan, C. M. (1977) Inheritance of selected pathotoxin resistance in maize plants regenerated from cell cultures. *Proc. Nat. Acad. Sci., U. S. A. 74*, 5113-17.

Guerrinean, M.; Slonimski, P. P.; and Avner, P. R. (1974) Yeast episome-oligomycin resistance associated with a small CC non-mitochondrial DNA. *Biochem. Biophys. Res. Communs. 61*, 462-69.

Hull, R. (1978) The possible use of plant viral DNAs in genetic manipulation in plants. *Trends in Biochem. Res. 3*, 254-56.

Kent, S. S.; Tingey, S.; and Young, J. D. (1979) Simultaneous kinetic analysis of RuBP carboxylase/oxygenase activities. *Plant Physiol.* (in press).

Kirk, J. T. O. (1971). Chloroplast structure and biogenesis. *Ann. Rev. Biochem. 41*, 161-96.

Laing, W. A.; Ogren, W. L.; and Hageman, R. H. (1974) Regulation of soybean net photosynthetic CO_2 fixation by the interaction of CO_2, O_2 and ribulose 1,5-diphosphate carboxylase. *Plant Physiol. 54*, 678-85.

Langer, I.; Frey, K. J.; and Bailey, T. B. (1978) Production response and stability characteristics of oat cultivars developed in different ears. *Crop Sci. 18*, 938-41.

Levings, C. S., and Pring, D. R. (1976) Restriction endonuclease analysis of mitochondrial DNA from normal and Texas cytoplasmic male sterile maize. *Science 193*, 158-60.

Lewis, W. H., and Lewis, P. F. E. (1977) *Medical Botany*, p. 515. Wiley Interscience Publishing Company, New York.

McClintock, B. (1967) Genetic systems regulating gene expression during development. *Develop. Biol. 1 (Suppl)*, 84-112.

Matern, U.; Strobel, G.; and Shepard, J. (1978) Reaction to phytotoxins in a potato population derived from mesophyll protoplasts. *Proc. Nat. Acad. Sci., U. S. A. 75*, 4935-39.

Miller, J. E. (1978) A proposal for the genetic manipulation of plant metabolism. *J. Theor. Biol. 74*, 153-54.

Moore, L. W.; Warren, G.; and Strobel, G. (1979) Involvement of a plasmid in the hairy root disease of plants caused by *Agrobacterium rhizogenes. Plasmid.* 2: 617-26.

Mujeeb, K. A.; Thomas, J. B.; Ridriguez, R.; Waters, R. F.; and Bates, L. S. (1978) Chromosome instability in hybrids of *H. vulgare* L. with *T. turgidum* and *T. aestivum. J. Heredity 69*, 179-82.

National Academy of Sciences. (1975) *Underexploited Tropical Plants with Promising Economic Value.* National Academy of Sciences, Washington, D. C.

National Research Council. (1977) *Supporting Papers: World Food and Nutrition Study*, vol. 1. National Academy of Sciences, Washington, D. C.

Okyama, K.; Gamborg, O. L.; and Miller, R. A. (1972) Uptake of exogenous DNA by plant protoplasts. *Can. J. Bot. 50*, 2077-80.

Princen, L. H. (1977) Potential Wealth in New Crops: Research and Development. In *Crop Resources* (D. S. Seigler, ed.) pp. 1-15. Academic Press, New York.

Schell, J., and Van Montagu, M. (1977) Transfer, maintenance and expression of bacterial Ti-plasmid DNA in plant cells transformed with *A. tumefaciens* in genetic interaction and gene transfer. *Brookhaven Symp. Biol. 29,* 36-49.

Shepard, J., and Totten, R. (1977) Mesophyll cell protoplasts of potato. *Plant Physiol. 60*, 313-16.

Strobel, G. A. (1974) Phytotoxins produced by plant parasites. *Ann. Rev. Plant Physiol. 25*, 541-66.

Strobel, G. A. (1977) Bacterial Phytotoxins. *Ann. Rev. Microbiol. 31*, 205-24.

Sullivan, W. (1979) Hope of creating perennial corn raised by a new plant discovery. *New York Times*, February 9, p. A9.

Van Larebeke, N.; Engler, G.; Holsters, M.; Van de Elsacker, S.; Zaenen, I.; Schilperoot, R. A.; and Schell, G. (1974) Large plasmid in *Agrobacterium tumefaciens* essential for crown gall-inducing ability. *Nature 252*, 169-70.

Genetic Needs of Plant Breeders

R. A. Kleese and D. N. Duvick

We want to say at the beginning that we appreciate the opportunity to tell you about some of our needs in plant breeding. We all have the genetic improvement of crop species as a common goal, and yet we bring together vastly different backgrounds. Our research experiences range from chemistry laboratories to farmers' fields, from working with microorganisms to higher plants. Some of you are fascinated by a molecular approach—others prefer to always work with whole organisms. Most of the journals that some of you read are unknown to the rest of us. Our scientific language varies so much that communication is often very difficult.

Because of these differences, we appreciate the opportunity to interact with the participants of this symposium. We believe this opportunity for communication is necessary for fostering new approaches to the genetic improvement of crop species.

Before dealing specifically with genetic needs of plant breeders, we would like to give some general background about where plant breeders spend most of their time and resources. This should help give a perspective on the complexity of the task of improving crop species.

Two Categories of Plant-Breeding Activities

We can conveniently separate plant-breeding activities into two categories: (1) manipulating genetic variability and (2) evaluation. Much of our effort goes into generating genetic variability for selection. Choices are made about which traits to combine as well as about the best genetic source of these traits. These selections are advanced to the stage at which they can serve either as varieties themselves or as parents in hybrid combinations. It is in this general area of manipulating genetic variability and selecting desired genotypes that we will focus most of our attention.

Field evaluation (yield testing) is the other major plant-breeding activity. The demands for space, equipment, and time often limit the size of the breeding program. In evaluating hybrids or varieties in the field, we encounter considerable environmental variation as well as an interaction between genotypes and the environment. Temperature, moisture, soil structure, and fertility; the use of herbicides and insecticides; previous cropping history; and various tillage systems are all parts of this environment. It is necessary to test over a wide range of these conditions to get the best prediction of what particular hybrids or varieties will do in succeeding growing seasons. The following are just a few examples of what we mean.

The yields of two corn hybrids in two years are given in Table 1. Most of the examples that we will use involve corn since that is the crop with which we are most familiar. However, the principles behind these examples can be extended to work with other crop species. Note that in 1976, 3709 yielded 9 bushels per acre more

Table 1. Variation in Hybrid Performance in Bushels per Acre in 1976 and 1978

Hybrid	1976	1978
3709	123**	124**
3780	114	136
Number of observations	209	73

**Hybrids significantly different at the 1% level of probability.

than 3780. In 1978, 3780 yielded 12 bushels more than 3709. Both of these hybrids are adapted to and grown extensively in southern Minnesota and northern Iowa. In 1976, southwestern Minnesota and northwestern Iowa faced a severe drought. This meant that much of the corn in the area was exposed to stress. We found in 1976, and have confirmed the observation since, that 3709 does relatively better than 3780 under stress conditions. In 1978, growing conditions in southern Minnesota and northern Iowa were excellent. With favorable growing conditions, 3780 is generally superior to 3709. There is a significant difference between these hybrids in their response to the environment.

The other observation to be made from Table 1 is the large number of comparisons between these two hybrids. Because of the tremendous environmental variation and the experimental error encountered in conducting these tests, it is necessary to make many comparisons.

In Fig. 1, we have given the average yield and the difference in yield of two hybrids when they are grown at a number of different locations. These two hybrids, 3732 and 3780, were grown in 1977 at 25 locations in northern Iowa, southern Minnesota, and southeastern South Dakota. The hybrids were grown side by side at each location. Each bar on the graph indicates the difference in yield between the two hybrids at a location. For example, the first bar at the base of the graph indicates that 3732 and 3780 yielded the same at that location. The second bar on the graph indicates that 3780 yielded 1 bushel per acre more than 3732 at the next location. The third bar indicates that 3732 yielded 13 bushels per acre more than 3780 at the third location, and so on through the 25 locations. In several locations, there were no differences in yield between the two hybrids. On the other hand, in thrcc locations 3732 yielded 13 bushels per acre more than 3780 and in one location 3780 yielded 19 bushels per acre more than 3732. The average difference between the two hybrids at these 25 locations was only 2 bushels per acre, in favor of 3732 (3732 with 131 bushels per acre and 3780 with 129).

There does not appear to be a significant relationship between yield level (that is, the average yield of both hybrids) and difference in yield. For example, both hybrids yielded the same at the loca-

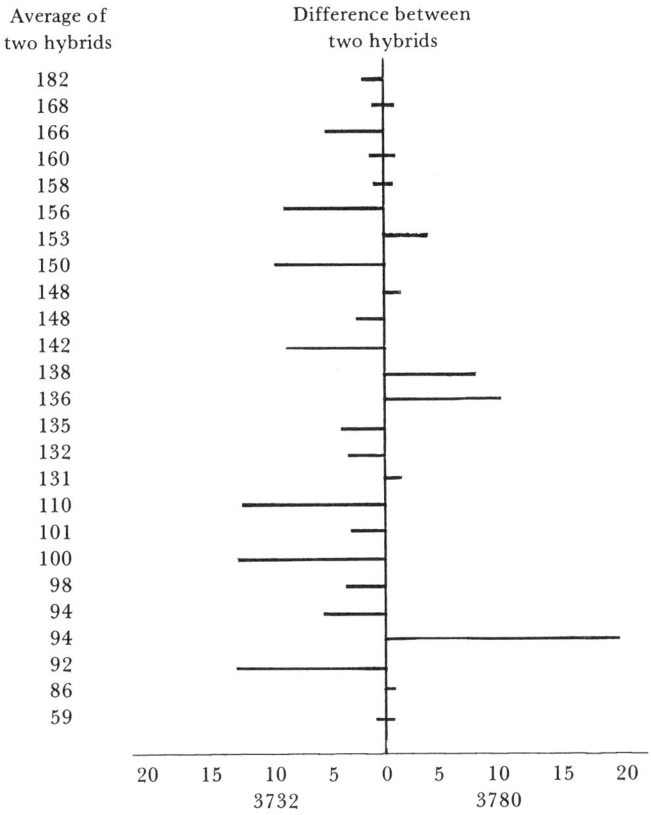

Fig. 1. Average yield (in bushels per acre) and difference in yield of two hybrids, 3732 and 3780, at 25 locations in 1977.

tion with the lowest average yield (59 bushels per acre) and at the location with next to the highest yield (168 bushels per acre). At the locations represented by the third and fourth bars from the base of the graph, the average yields were nearly the same; yet the differences in yield were very large, favoring 3732 in one case and 3780 in the other.

Evaluation is not simply a matter of contending with year-to-year or location-to-location variability. One additional factor of interest, when corn hybrids are compared, is their performance at varying population densities. Population density refers to the number of plants per unit area. We have two reasons for being interested in this kind of evaluation. First, farmers vary the number of plants that they establish per acre. Farmers who anticipate low moisture or fertility may choose to plant a low number of plants per acre; others who expect good moisture and fertility may choose a medium or high population density. The other reason for being interested in making evaluations under varying population densities is to gather data on the ability of the hybrid to perform under stress conditions. If the amounts of available moisture or nutrients become limited during the growing season, plants grown at high population density may be severely stressed.

Fig. 2 shows examples of two kinds of response to increasing population density. The hybrid 3382 increases in yield as the population density increases. In contrast, 3388 is nearly the same at the low and medium rates and then decreases at the high rate. The statistical interaction of hybrid multiplied by population density, a measure of the differential response of hybrids to varying population density, is significant at the 1% level.

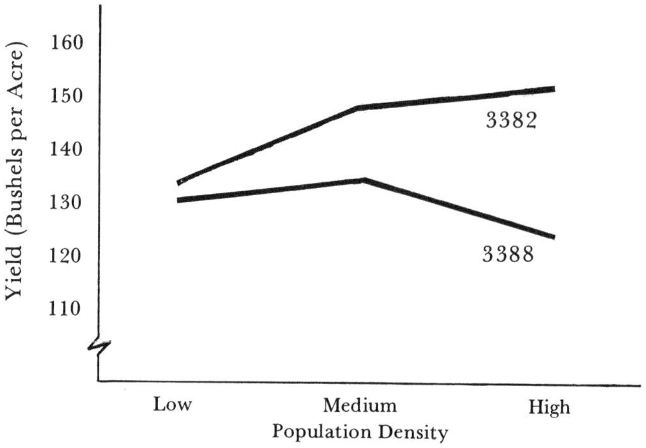

Fig. 2. Yield versus population density of two hybrids in 1977. Hybrids differed at the 1% level, populations differed at the 5% level, and their interaction differed at the 1% level of statistical probability.

In the northern Corn Belt, we are interested in hybrid maturity because we must know which hybrids mature before the killing frosts in the fall. This becomes less important in choice of hybrids as one moves south in the Corn Belt. Because of the livestock and cropping enterprises on his farm, a farmer in southern Iowa may choose to grow a hybrid that matures much earlier than the anticipated date of the first killing frost. For this reason, we are interested in evaluating hybrids south of the area (maturity zone) where they would be considered full-season hybrids. Table 2 shows a comparison between two hybrids evaluated in four different maturity zones. A maturity zone covers approximately 50 miles from north to south and is largely a function of temperature. Zones 3 through 6 generally are located in southern Minnesota, northern Iowa, central Iowa, and southern Iowa, respectively. In zones 3, 4, and 5, hybrids 3732 and 3780 have comparable yields; however, in zone 6, 3780 is significantly better than 3732. There is a marked difference between these hybrids in their response to the length of the growing season.

The point we wish to make with the above examples is that field evaluation, or yield testing, is not a simple matter of picking the best hybrid. We must test so that we take into account a wide range of environmental conditions (including temperature, moisture, and soil) as well as the wide range of farmers' cultural practices and needs that fit their particular farming enterprises. We spend a lot of time, money, and manpower on these evaluation activities and, as was stated previously, they often limit the size of our breeding projects.

Table 2. Yield (in Bushels per Acre) of Two Hybrids versus Maturity Zone in 1977

Hybrid	Zone 3	Zone 4	Zone 5	Zone 6
3732	147	140	135	117*
3780	144	142	135	128
Number of observations	99	108	30	18

*Hybrids significantly different at the 5% level of probability.

Genetic Needs as Seen by Plant Breeders

With this background, let us turn now to particular genetic needs of plant breeders. In an effort to assess these needs across a broad spectrum of crops and plant breeders, we surveyed the plant breeders within Pioneer Hi-Bred International, Inc. We told each of the plant breeders that we were preparing a presentation on the genetic needs of plant breeders and would like their help in evaluating these needs. They were asked to list the 10 most important genetic needs in *their* breeding programs and to rate the relative importance of each on a scale of 1 to 9, with 9 being most important. We received 42 responses from plant breeders working on corn, sorghum, soybeans, wheat, cotton, and alfalfa. Since most of our plant breeders work with corn, most of the responses were oriented to this species.

Genetic Need: Pest Resistance

The nine most important traits as judged by the survey are given in Table 3. We have calculated "relative importance," which is simply the product of the number of responses and the average rating of importance. The genetic need with the highest score is resistance to pests. This category includes resistance to a number

Table 3. Nine Most Important Genetic Needs as Judged by Plant Breeders

Genetic Need	Relative Importance[a]
Resistance to pests	422
High, stable yield	370
Resistance to stress conditions	212
Utilization of new germ plasm	199
Maturity alteration	124
Genetic control of negatively correlated traits	122
Efficient methods of selection	115
Genetic control of individual morphological traits	115
Good standability	112

[a]Product of number of responses multiplied by average rating. Highest numbers are most important.

of different diseases and insects. We have needs for efficient methods of incorporating resistance into otherwise desirable genotypes, for greater genetic diversity of sources of resistance, and for a better understanding of the genetic basis of single-gene and multigene resistance. We need to understand the genetic basis of antibiosis.

It is worth considering in more detail the distribution of some diseases and insects that affect corn. Southern corn leaf blight, *Cochliobolus heterostrophus*, is a fungal disease found primarily in the mid-South (Kentucky and Tennessee) and the southeastern United States. Generally, it is not of consequence in the central Corn Belt, although race T of this disease did move up into Iowa, Illinois, and Indiana and cause extensive damage in 1970. This forced a rapid change to resistant hybrids that has virtually eliminated race T.

Downy mildew, *Sclerospora sorghi*, can be found on corn and sorghum, generally in the area where sorghum is grown (from Kansas to Texas). This pathogen often attacks early in the plant's life with the result that no grain is produced.

Virus diseases of corn have generally been concentrated in the Ohio River Valley and the South. They usually have not been significant in the southeastern United States. Maize dwarf mosaic virus and maize chlorotic dwarf virus are two of the more important viruses found in the southeast. Field identification of viral diseases is very difficult. Problems arise because of the occurrence of many different viruses and strains of viruses, indistinct symptomatology, and confusion with nonpathogenic kinds of plant abnormalities.

Northern leaf blight, *Helminthosporium turcicum*, occurs wherever corn is grown in the United States. It can be severe in the central Corn Belt as well as on winter-grown crops in Florida and Texas. Single-gene resistance has been incorporated into a large number of inbred lines.

Diplodia (*Diplodia maydis*), Gibberella (*Gibberella roseum*), and Fusarium (*Fusarium moniliforme*) are the major stalk rot pathogens. Stalk rot usually occurs in the lower portion of the stalk, causing it to weaken and eventually fall over. Stalk rot is an important disease throughout the corn-growing area and is one of the most destructive diseases of corn.

The European corn borer, *Ostrinia nubilalis*, can be found in most areas where corn is grown. The first brood of larvae feed on leaves before they crawl down behind leaf sheaths and into the stalk to pupate. Moths later emerge to produce a second generation of larvae. This generation can be very destructive, boring into stalks, ears, and ear shanks as well as removing leaves from the plants. Because of the borers' feeding, stalks and ear shanks are often so weakened that a slight breeze will cause the stalks to fall over or the ear shank to break off.

The southwestern corn borer, *Diatraea grandiosella*, is generally confined to the mid-South. This insect girdles the stalk just above the ground, causing the stalk to fall over.

Many diseases and insects we have not mentioned here are destructive to corn (Ullstrup, 1976; Dicke, 1976). The major points to be made are that disease and insect problems vary in kind and severity from one region to another. The breeder must continually be alert to new disease and insect problems and to new strains of existing pathogens. In some cases, single-gene resistance will be useful in combating the pests, but more often resistance will be genetically complex and not conveniently manipulated into new lines. Also single-gene resistance often lasts for only a few years because new strains of the pest arise. Thus, multigene resistance is usually the most desirable kind. See Day (1974) for a comprehensive treatment of the genetics of resistance to plant pests.

Genetic Need: High, Stable Yield

The next most important need is for high, stable yield. This need includes high seed yield in grain crops as well as high forage yield in legumes. There is a need for greater stability of yield across environments so that a hybrid or variety can be moved from one region to another without losing high yield. There is a need for a better understanding of the genetic control of the factors that limit yield, such as plant architecture and photosynthetic rates. What is the net value of symbiotic nitrogen fixation to yield? We need to know more about the genetic control of the distribution of carbohydrates to grain and leaves; the genetic control of the abortion of flowers, seeds, and pods; and the mechanism of photorespiration in C_3 species.

Genetic Need: Resistance to Stress

We need resistance to stress conditions. There is a need to better understand the physiological basis for heat and drought resistance or tolerance and the genetic control of these processes. We need to better understand what is involved in efficient water use and the relationship of stomatal number and leaf surface wax to water use and their genetic control. We need to understand the control of metabolism at low water potential. Are certain morphological traits associated with particular types of stress tolerance? Can we use these traits for selection, or must we depend on cumbersome and lengthy laboratory procedures?

Genetic Need: Efficient Incorporation of New Germ Plasm

Incorporating new germ plasm is an important genetic need. We should develop elite populations as source materials for future breeding. How do we decide which materials to include in these populations? How should we generate the populations and how should we handle them from one selection cycle to the next? How can we best recover the unique gene combinations that we desire? How do we incorporate exotic germ plasm into elite without bringing in undesirable linked genes? How can we ensure a broad genetic base for our elite material?

Genetic Need: Control of Maturity

There is a need to better understand the genetic control of maturity. What do we know of the genetic control of photoperiod sensitivity, temperature sensitivity, vernalization, and winter hardiness? What is the genetic signal that ends grain filling in a cereal? Are there specific genes with major effects on maturity that would enable us to move a hybrid from the southern to the northern Corn Belt?

In Table 4 are data that pertain to the different rates of growth, development, and maturity of three different corn hybrids. Growing degree units (GDUs) is a system of measuring units of heat during the corn-growing season. There is a strong positive correlation between GDU and the growth and development of corn. GDU to silk indicates the number of units required for a particular hybrid to grow from germination to the time at which the silks appear. GDU to black layer (BL) is the number of units of heat required for the hybrid to grow from germination to physiological maturity, that is, the time at which dry-matter accumulation in the grain has ceased. Minnesota Relative Maturity (RM) is an index used to define the relative length of the growing season needed by a particular hybrid to reach a harvestable grain-moisture percentage.

The three hybrids listed in Table 4 all have the same RM and are considered to have the same maturity based on grain-moisture percentage at harvesttime. If we examine the GDUs to silk and the GDUs to black layer, we can see that these hybrids proceed through their growth and development at different rates. Hybrids 3732 and 3780 silk about 40 units earlier than 3728. The silking period would occur in early July, and 40 units would represent about a two-day difference between the hybrids. Although 3780

Table 4. Length of Grain Fill of Three Corn Hybrids of the Same Maturity

Hybrid	GDU Silk	GDU BL	RM
3728	1346	2516	107
3732	1300	2454	107
3780	1307	2557	107

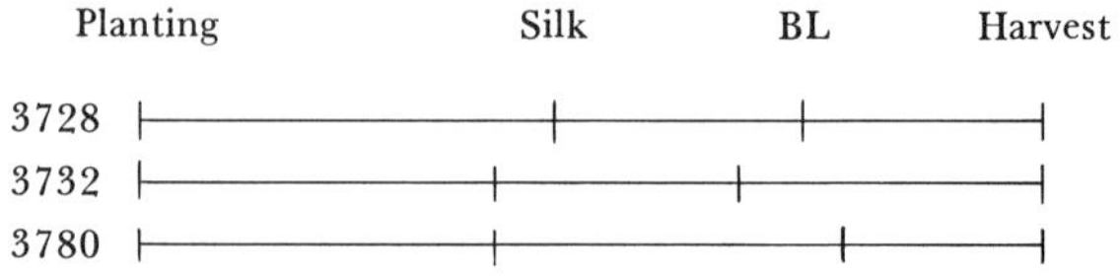

Fig. 3. Relationship of grain-filling periods of three corn hybrids of same maturity.

silks earlier than 3728, it does not reach black layer (that is, physiological maturity) until after 3728. On the other hand, 3732 is early to silk and early to black layer. The time lines in Fig. 3 may more clearly illustrate the differences among the hybrids when they are grown side by side in southern Minnesota.

The period of time from flowering to physiological maturity is the time during which the grain is filling. If the rates of grain fill from the three hybrids were comparable, we would obviously prefer a hybrid such as 3780 since it spends considerably more time in this stage of development. However, as some of the data in Table 2 show, the yield of 3732 was equal to or slightly better than 3780 when the hybrids were grown in southern Minnesota and northern Iowa. Therefore, the rate of grain fill for 3732 must have been greater than for 3780 in that environment.

One further point is worth mentioning. The moisture contents of all three of these hybrids at physiological maturity are higher than at harvest maturity and are about equal for the three hybrids, as we know from another set of data. Basically, the only change that takes place in the grain from physiological maturity to harvest is the loss of water. If all three of these hybrids could be harvested at the same time (that is, if they had the same grain moisture content at harvest time) the rates of drying would be different among the three hybrids since each has a different length of time from BL to harvest (see the time lines in Fig. 3). In fact, we know that 3780 dries more rapidly than the other two hybrids.

If possible, we would prefer to have the best of all combinations, that is, a long grain-filling period, a high rate of grain fill, and a rapid drydown. If we attempt to alter maturity, we may change any or all of these periods of growth and development. Today, we cannot easily manipulate these three periods of development independently.

Genetic Need: Manipulation of Negatively Correlated Traits

How can we manipulate negatively correlated traits? We are often forced to a compromise between a good trait and a poor trait, for example, high yield and poor stalks or high yield and low protein. If we get strong stalks, they may be brittle. Some sources of disease or insect resistance, such as greenbug tolerance in sorghum, often bring in low yield. The semidwarf habit in wheat may be linked with factors for brittle straw and low test weight. We need to better understand the physiological basis of these traits and their genetic control so that we might be able to manipulate them simultaneously to our advantage.

Genetic Need: Efficient Ways to Select in Segregating Generations

Plant breeders are always interested in more efficient methods for selecting in segregating generations. During the early generations, which traits should we emphasize? If a hybrid is the end product, should we evaluate the unfinished lines per se or in hybrid combinations? If hybrid combinations are used, how many testers and what kinds of testers should we use? How early in the segregating generations should evaluation begin?

Genetic Need: Standability

Standability is an important trait in all grain crops. The crop must stand well enough at harvest time so that harvest is efficient and losses are minimal. What is the role of root strength and morphology in supporting the plant and how can we alter these traits genetically? How important is mechanical strength in good stalks versus resistance to stalk rot organisms?

Genetic Need: Control of Morphological and Anatomical Traits

The final category of genetic needs asks for a better understanding of the genetic control of particular morphological and anatomical traits. What is the genetic basis for plant and ear height, ear retention, ear and kernel size, and the various factors involved in grain quality?

Table 5. Plant Breeders' Ratings of six traits Often Suggested as Amenable to Laboratory Manipulation

Genetic Need	Relative Importance[a]
Altering level of heterozygosity/homozygosity	47
Genetic control of breeding behavior	44
Genetic control of nutritional quality	39
Efficient uptake and utilization of nitrogen	32
Evaluation of the potential for tissue culture and genetic engineering	27
Resistance to herbicides	12

[a]Product of number of responses multiplied by average rating. Highest numbers are most important.

Genetic Need: Controlled Heterozygosity or Homozygosity

In Table 5 are listed those traits that are often suggested as possibilities for manipulation in the laboratory. The relative importance of these needs is significantly less than those listed in Table 3. The first need listed is the ability to manipulate the level of homozygosity or heterozygosity. This could be of value in furthering our understanding of the genetic basis of heterosis. It might also lead to ways in which we could maximize heterozygosity. Some breeders of self-pollinated species would like to develop homozygous lines quickly.

Genetic Need: Controlled Breeding Behavior

There is a need to control the breeding behavior in several crop species. Additional sources of cytoplasmic male sterility would be desirable in any crop species in which this form of male sterility is used to produce hybrids. We need a better understanding of the genetic control and biochemical basis of fertility restoration as it is used in combination with cytoplasmic male steriles. We know very little about the genetic basis of apomixis.

Genetic Need: Ways to Alter Nutritional Quality

We have several good examples of genetic alteration of nutritional quality in crop species. Alterations of lysine content in barley, corn, and sorghum grain; starch and sugar composition of corn endosperm; and protein content in corn, oats, and wheat are several of the better known examples. Yet, problems result when some of these changes are made. How can we significantly improve lysine content in corn without adversely affecting grain quality and yield? How can we be more effective in improving protein content and yield simultaneously? We know very little about the genetic control of the various parameters of forage quality in forage species.

Genetic Need: Control of Nitrogen Uptake and Utilization

There is a need to better understand the genetic control of uptake and utilization of nitrogen in crop species. What is the metabolic cost of symbiotic nitrogen fixation? Can we develop varieties of legumes that are more responsive to infection with highly efficient strains of rhizobia?

Genetic Need: Understanding the Use and Potential of Genetic Engineering for Plants

How do we assess the potential of tissue culture and genetic engineering to improve crop production? Where should the emphasis be? What kinds of traits should be studied?

Genetic Need: Knowledge about Herbicide Resistance

What do we know of the genetics and physiology of resistance to herbicides? As we become more oriented toward chemical cultural practices, will this become a more significant issue?

Genetic Needs Usually Are for Genetically Complex Traits

As we look over the traits listed in Tables 3 and 5, it is apparent that most of the genetic needs are not simple or clearly defined from a biochemical or physiological standpoint. There are examples of single genes that control resistance to disease, alter photoperiod sensitivity, control plant height, control breeding behavior with nuclear male sterile genes, control some types of nutritional quality (such as high lysine in corn and sorghum and starch composition in corn and barley), and control resistance to some herbicides. We might also add cytoplasmic male sterility to this list. We do not understand it as a single-gene phenomenon, but we have been able to manipulate it rather easily with routine breeding procedures.

Although single-gene control of some traits does exist, there may be reasons for preferring more complex genetic systems (Day, 1974; Horsfall, 1972). For example, a common approach in breeding for disease resistance has been to search for single genes that govern resistance to a particular race of a pathogen. These genes were incorporated into agronomically desirable lines. In time, the pathogen altered itself sufficiently so that new virulent races were produced. At this point, a new search was made for resistance genes and the cycle was repeated. Should breeding efforts move more in the direction of incorporating genetically more complex forms of resistance that theoretically might make it more difficult for the pathogen to generate the new virulent form? Should—or could—the complexity be at the single-gene level, or should multiple genes be used?

Can Traits Be Reduced to Simple Components for Genetic Manipulation?

In general, the most important genetic needs are genetically complex. Could we reduce the complex traits to their less complex components? Would this approach allow us to better understand and genetically manipulate the more complex traits?

There are several good examples of how plant breeders and physiologists have pursued this line of investigation (component analysis) (Frey, 1971; Moss and Musgrave, 1971; Wallace et al., 1972). Morphological yield components are a well-known example of this component approach to plant breeding. Basically, the morphological components of grain yield on a per-plant basis would be the number and size of seeds. If one were dealing with corn, the number of seeds could be broken down into two components, the number of ears per plant and the number of seeds per ear. Additionally, one could deal with components of seed size, that is, length, width, and density. In fact, there are always components of the components. At some point, these components may become anatomical, physiological, or biochemical components.

One important physiological component of yield would be rates of photosynthesis. But to consider this component it would seem necessary to ask what the net photosynthetic rate is. How much photosynthate is respired and, therefore, reduces the amount left to be used for further growth or for storage in the grain? How do the levels and specific activities of the enzymes essential to photosynthesis affect the overall rate? What effect does stomatal number and size have on carbon dioxide diffusion through the leaf? What effect does leaf angle, area, and thickness have on light interception and thus on photosynthetic rate? What effect does translocation of photosynthate away from the chloroplast have on photosynthetic rate?

It is not surprising that attempts to correlate any of these components with yield generally have been unsuccessful. Any one of these factors could be the limiting factor for the yield of any particular variety or hybrid. However, there would be a unique combination of physiological and morphological components of yield for each genotype. A question to ask is whether any one of these components is found to be limiting yield much more often than any of the others. It is sufficient to say at this point that the component approach to breeding for yield has not greatly altered breeding procedures. Rather, the component approach causes the breeder to alter his opinion about what is desirable as he moves through his nursery.

Today, the corn breeder probably pays more attention to erect leaf habit than he did in the past, although he certainly would not throw away material that is good in several other traits but simply does not have an erect leaf habit. He considers the length of the ears and the number of kernel rows, although in most cases he does not go to the trouble to measure these attributes. He considers the size of the tassel. He would like to have it shed enough pollen so that he can easily maintain the line and perhaps use it as a male parent. On the other hand, he may be concerned that a very large tassel could be a detriment to the overall productivity of the line. These are subtle matters of judgment, and no two plant breeders would judge them in the same way.

In general, if a plant breeder is interested in yield, he still measures the yield itself rather than the various components of yield. The use of yield components in breeding for higher yields should not at this time be considered an unsuccessful approach. Rather, we need to know more—much more—about the physiological basis of yield and its genetic control.

What we have said about yield and yield components can be said about the other complex genetic traits that are important to improving crop plants. One can define components of standability, of maturity, and of resistance to stress. We need very much to study the physiological processes involved in these traits and their genetic control.

Manipulating Crop Plants in the Laboratory

With this as background, what can we suggest as possible uses of laboratory approaches for improving crop plants? There is a need for manipulating those traits that are controlled by individual genes. Examples of this are male sterility, some kinds of nutritional quality, and some kinds of disease, insect, and nematode resistance. But we need a good, concrete example of a single gene from a higher plant being genetically engineered into another plant. Instead of trying to manipulate these traits, it might be better, at least from an identification standpoint, to work with a gene whose expresssion is easily measured, such as an enzyme activity or plant or seed color. This could teach us a great deal about the mechanics of accomplishing this transfer as well as giving impetus to greater research activity in this field.

Could laboratory approaches assist us in increasing recombination in certain chromosomes or selected parts of chromosomes while maintaining blocks of desired genes? Could we recombine genes without going through the sexual cycle? Is there a laboratory system that would allow us to make "crosses" or "selfs" at any time of the year, independent of the restriction of temperate latitudes, greenhouses, and winter nurseries? Can we manipulate levels of heterozygosity or homozygosity more precisely?

Need for Basic Knowledge of Plant Biochemistry Physiology, and Development

Perhaps our greatest need, both with laboratory and field approaches to crop improvement, is to learn much more about how plants grow, develop, and reproduce. This basic information must be on hand before we can make efficient use of the much-publicized genetic-engineering techniques. We need to identify gene function of many more genes and, in particular, of those genes that have importance for productivity in the crop species. The book *The Mutants of Maize* (Neuffer et al., 1968) lists 259 genes; one can find practical possibilities for only about 10% of these genes. We need much more information about various metabolic pathways and their genetic control in plants. We know almost nothing about genetic regulation in plants. We need to study the genetics of plant regeneration from callus. We know very little about the genetic control of hormone level in plants or, conversely, of the hormonal control of gene expression. We know almost nothing of genetic control of recombination.

Plant breeding can be a very complex business, and, because of this, we need as many tools as we can obtain. We should not be deterred by this complexity but should *choose the most simple, straightforward problems for initial genetic-engineering efforts.* We think that, in attacking simple problems, we will learn a great deal more about what we can and cannot expect to accomplish with some of the laboratory approaches to plant improvement.

Plant breeders have a tremendous array of genetic stocks and of ways to evaluate new genotypes for practical use. We need to communicate with scientists working in the area of genetic engineering to make available some of these unique genetic materials and then to help evaluate the usefulness of the genetic-engineering products. The greater the interaction between these two groups, the greater the opportunity for improving crop species.

References

Day, Peter R. (1974) *Genetics of Host-Parasite Interaction,* p. 225. W. H. Freeman, San Francisco.

Dicke, F. F. (1976) The most important corn insects. In *Corn and Corn Improvement* (G. F. Sprague, ed.), pp. 501-90. American Society of Agronomy, Madison, Wisconsin.

Frey, Kenneth J. (1971) Improving crop yields through plant breeding. In *Moving Off the Yield Plateau* (Jerry D. Eastin and Robert D. Munson, eds.), pp. 15-58. American Society of Agronomy, Madison, Wisconsin.

Horsfall, James G., chairman, Committee on Genetic Vulnerability of Major Crops. (1972) *Genetic Vulnerability of Major Crops,* p. 304. National Academy of Sciences, Washington, D.C.

Moss, Dale N., and Musgrave, Robert B. (1971) Photosynthesis and crop production. *Adv. Agron. 23,* 317-36.

Neuffer, M. G.; Jones, Loring; and Zuber, Marcus S. (1968) *The Mutants of Maize,* p. 74. Crop Science Society of America, Madison, Wisconsin.

Ullstrup, A. J. (1976) Diseases of corn. In *Corn and Corn Improvement* (G. F. Sprague, ed.), pp. 391-500. American Society of Agronomy, Madison, Wisconsin.

Wallace, D. H.; Ozbun, J. L.; and Munger, H. M. (1972) Physiological genetics of crop yield. *Adv. Agron. 24,* 97-146.

Part II: Genome Organization and Function

Patterns of DNA Sequence Repetition and Interspersion in Higher Plants

W. F. Thompson, M. G. Murray, H. S. Belford, and R. E. Cuellar

It is natural to think of DNA in terms of genes and their controlling elements. However, if genes and controlling elements actually made up a large fraction of the total DNA, we should expect that the DNA content of different organisms would be rather directly correlated with organismic complexity; in fact, the correlation is very far from perfect. Although the *minimum* DNA content for various major taxa does increase in a more or less regular fashion with increasing complexity (Sparrow et al., 1972), an even more striking feature of the distribution of DNA contents among organisms is the tremendous range of variation encountered even among those with similar biological complexities. For example, animals usually contain between 1 and 5 pg of DNA per haploid genome, but the range for salamanders extends to 100 pg or so,

Note: This work was supported in part by grants from the National Science Foundation (PCM 7705656 and DEB 76-83405) and from the Competitive Grants Office of the United States Department of Agriculture (78-00613). R. E. Cuellar received support from a Ford Foundation predoctoral fellowship. We are grateful to Glenn Ford for expert assistance with the computer programming and to Dr. Winslow R. Briggs for not making critical comments on the manuscript. This paper is publication number 659 of the Carnegie Institution of Washington, Department of Plant Biology.

and primitive vascular plants may contain as much as 300 pg (Hinegardner, 1976). Even among angiosperm species, we find a nearly 100-fold range of variation and significant differences are seen even within such restricted taxomic groups as genera.

The large and variable sizes of eukaryotic genomes in general, and those of plants in particular, suggest that much of the "genetic material" in many of these organisms is not needed to account for known coding or regulatory functions in development. What, then, is all this "excess" DNA doing in so many genomes? How did it get there? Could it really be useless or functionally redundant, or might it serve some purpose that we have not yet perceived? In this paper, we review some of the basic data on the overall structure of higher plant genomes, with particular emphasis on our own work with peas and mung beans. This information provides a basis upon which we can begin to speculate about these questions. However, even more importantly, we think it also brings us face to face with the magnitude of the problems we must solve if we are to proceed beyond speculation to real understanding.

DNA Reassociation

Many of our current concepts of genome organization are derived from DNA-reassociation experiments in which the two component strands of double-helical DNA are first separated (denaturation) and then complementary strands are allowed to re-form base-paired structures during incubation under carefully defined conditions of salt and temperature. The major steps in a typical experiment are outlined in Fig. 1.

Three techniques are commonly used to measure the progress of reassociation. Since strand separation produces a hyperchromic shift that is reversed as base pairs re-form, reassociation may be followed by monitoring the decrease in optical density (hypochromic shift) as a function of time. Extremely precise measurements can be made this way, and the technique is nondestructive; a single sample of DNA can be monitored continuously over a long period of time. The range of concentrations that can be employed in optical experiments is limited, however, and hypochromicity

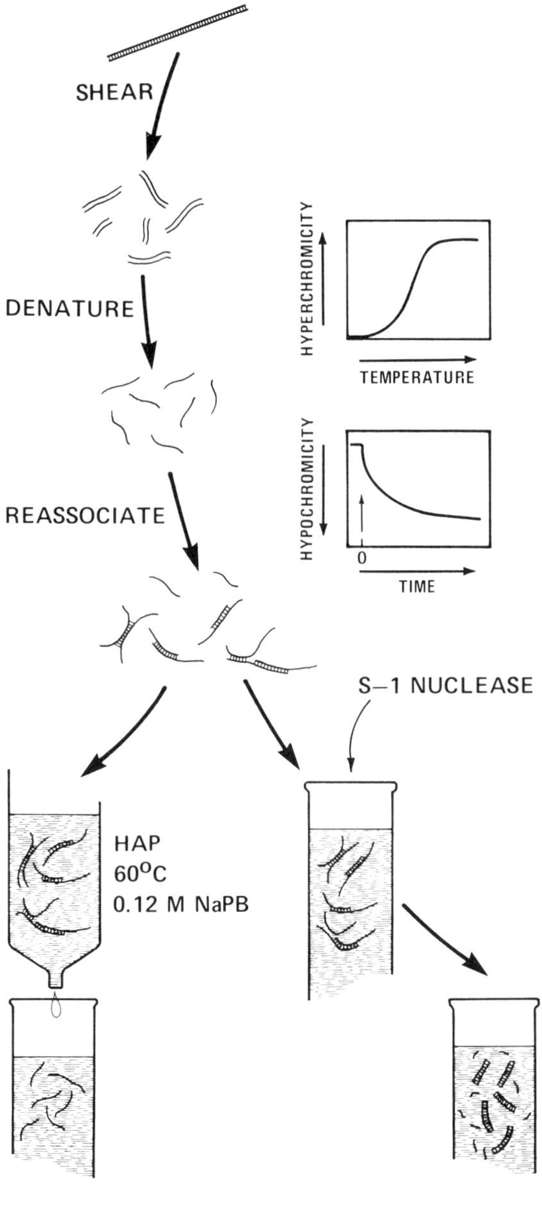

Fig. 1. Some essential steps in reassociation experiments. (From Thompson and Murray, 1979; by permission.)

measurements do not permit one to follow the behavior of different DNA samples in a mixture, as is possible when tracers are used in other assay systems. Two assays suitable for use with tracers are nuclease digestion and hydroxyapatite (HAP) chromatography. In nuclease assays, a single-strand-specific endonuclease such as *S-1* is used to digest the single-stranded DNA that remains in aliquots of the reassociation mixture sampled after various periods of incubation. HAP chromatography assays are based on the ability of hydroxyapatite to bind double-stranded DNA more tightly than single strands. Under appropriate conditions, single strands will pass through the column while double strands are retained.

Both optical and nuclease assays measure the pairing of individual nucleotides. In contrast, HAP chromatography measures the fraction of DNA *fragments* that have formed a duplex over any portion of their length since the single-stranded regions of such fragments will be retained on the column by virtue of their attachment to the tightly bound duplexes. The minimum duplex length required for binding is usually considered to be about 50 nucleotide pairs (Wilson and Thomas, 1973).

When assayed by HAP chromatography, reassociation of simple DNA follows essentially ideal second-order kinetics, as expected for a bimolecular reaction. Under a given set of reassociation conditions, the reassociation rate for *nonrepeated* DNA is inversely proportional to the genome size (Britten and Kohne, 1968). This relationship can best be understood in terms of the concentration of each individual sequence in the total population. As a genome size increases, the concentration of each sequence in a given amount of total DNA decreases, and so the frequency of collisions between complementary sequences, and hence the rate of reassociation, also decreases.

Repetition Frequencies

In addition to nonrepetitive or single-copy sequences, eukaryotic genomes typically contain substantial amounts of DNA that reassociate much more rapidly than expected for single-copy DNA and thus appear to be composed of sequences present in a number of copies in each haploid chromosome set (Britten and Kohne, 1968). Given an accurate measurement of the reassociation rate for a group of these *repetitive sequences*, it is possible to calculate an apparent repetition frequency by comparing that rate with the rate for single-copy sequences in the same genome. When the observed reassociation kinetics are complex, indicating the presence of more than one frequency class, it is usual to analyze the data with a computer program (e.g., Pearson et al., 1977) designed to fit a number of ideal second-order components to the data. Table 1 summarizes the results of such fitting procedures for DNA from a number of higher plant species. It is immediately apparent that the range of repetition frequencies encountered is quite broad, although, in most cases, components accounting for a significant fraction of the total DNA seem to be present in a few thousand copies or less.

Table 1 also lists estimates of the *kinetic complexity* for each repetitive component. Kinetic complexity is defined (Britten et al., 1974) as the total length of DNA contained in *different* nucleotide sequences, it is determined by comparing the reassociation rate of an unknown sample to that of a known nonrepetitive standard. Such measurements provide an index of information content that is not affected by total mass or repetition frequency. Repetitive sequence components can have relatively high kinetic complexities and thus potential information contents; in some cases, the estimates reach values of several million nucleotide pairs, or roughly the size of a bacterial genome. However, components with much lower complexities also exist. For example, about 3% of soybean DNA has been reported to be composed of sequences repeated about 300,000 times, with a kinetic complexity of only about 200 nucleotide pairs (Goldberg, 1978).

Although the data in Table 1 provide a useful indication of the range of repetition and complexity to be found among higher plant DNAs, it is important to recognize that the values given are only crude approximations. For example, the members of a typical repetitive-sequence family exhibit considerable sequence divergence so that they no longer reassociate with one another to produce precisely paired duplexes. In many cases, the average amount of mispairing observed in duplexes formed under standard conditions

Table 1. Component Analysis of Repetitive DNA Reassociation Kinetics

Species and Reference	Component	Fraction	Copies	Kinetic Complexity (NTP)
Vigna radiata (mung bean)	Very fast	0.02	–	–
Murray and Thompson, 1979	Fast	0.05	10,000	2.8×10^3
	Slow	0.28	175	9.0×10^5
Gossypium hirsutum (cotton)	Very fast	0.08	–	–
Walnot and Dure (1976)	Repetitive	0.27	130	1.6×10^6
Glycine max (soybean)	Very fast	0.05	–	–
Embryo DNA, Gurley et al. (1979)	Fast	0.28	3850	9.4×10^4
	Slow	0.24	153	2.0×10^6
Leaf DNA, Goldberg (1978)	Very fast	0.07	–	–
	Fast	0.28	2784	2.2×10^5
	Slow	0.38	19	1.8×10^7
Petroselinum sativum (parsley)	Very fast	0.05	–	–
Kiper and Herzfeld (1978)	Highly repetitive	0.13	136,000	1.8×10^3
	Fast	0.48	3000	3.0×10^5
	Slow	0.16	42	7.4×10^6
Nicotiana tabacum (tobacco)	Very fast	0.02	–	–
Zimmerman and Goldberg (1977)	Fast	0.06	12,400	4.6×10^3
	Intermediate	0.59	252	2.5×10^6
Pisum sativum (pea)	Very fast	0.04	–	–
Murray et al. (1978)	Fast	0.46	10,000	2.0×10^5
	Slow	0.33	3000	5.0×10^6

is of the order of 10%. This amount of mismatch has been shown to reduce the reassociation rate by a factor of two or more (Bonner et al., 1973; Marsh and McCarthy, 1974), and much greater effects would be anticipated for duplexes with greater than average mismatch. In addition, the division of complex genomes into a small number of discrete frequency classes is frequently quite arbitrary and artificial. C_0t curves for many plant DNAs are nearly linear throughout the range in which repetitive sequences are reassociating; this is illustrated for pea DNA in Fig. 2. Although computer fitting procedures can usually model these curves with between one and three ideal second-order components, the mere construction of such models does not constitute evidence that their components actually exist as physically discrete entities.

Relatively few attempts have been made to demonstrate the reality of discrete components in plant DNA by direct physical isolation, and those that have been reported usually failed to pro-

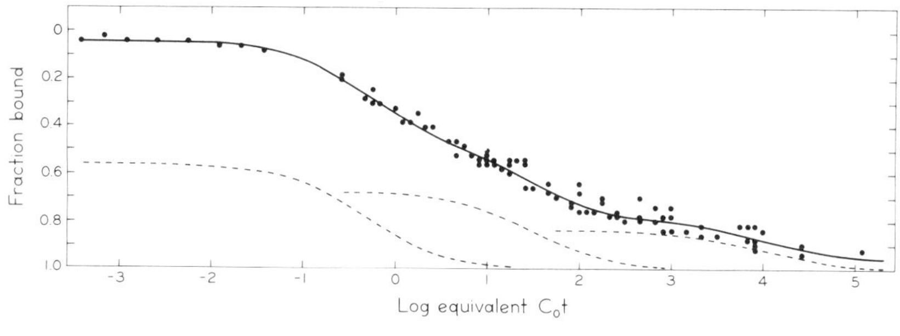

Fig. 2. Reassociation kinetics of short pea DNA fragments. DNA sheared to a modal single-strand length of 300 nucleotides was denatured and reassociated at different concentrations for various periods of time. The fraction of fragments bound to HAP (i.e., containing duplex DNA) at each point in the reaction is plotted as a function of C_0t, where C_0 is the DNA concentration (nucleotide molarity) and t is the time of reassociation (seconds). C_0t values obtained in different buffers have been converted to equivalent C_0t values for standard conditions, according to Britten et al. (1974). The dashed lines represent theoretical second-order components, of the best three-component model describing the data. According to this model, 16% of the fragments reassociate as a slow component corresponding to single-copy DNA, while 32% and 44% reassociate as though they contain sequences repeated 300 and 10,000 times per haploid genome, respectively. (Murray et al., 1978.)

duce convincing evidence. A DNA fraction whose reassociation approximates that expected for a single, ideal component has been obtained from tobacco DNA (Zimmerman and Goldberg, 1977). However, further analysis of an apparently simple repetitive component in mung bean DNA has shown it to be composed of at

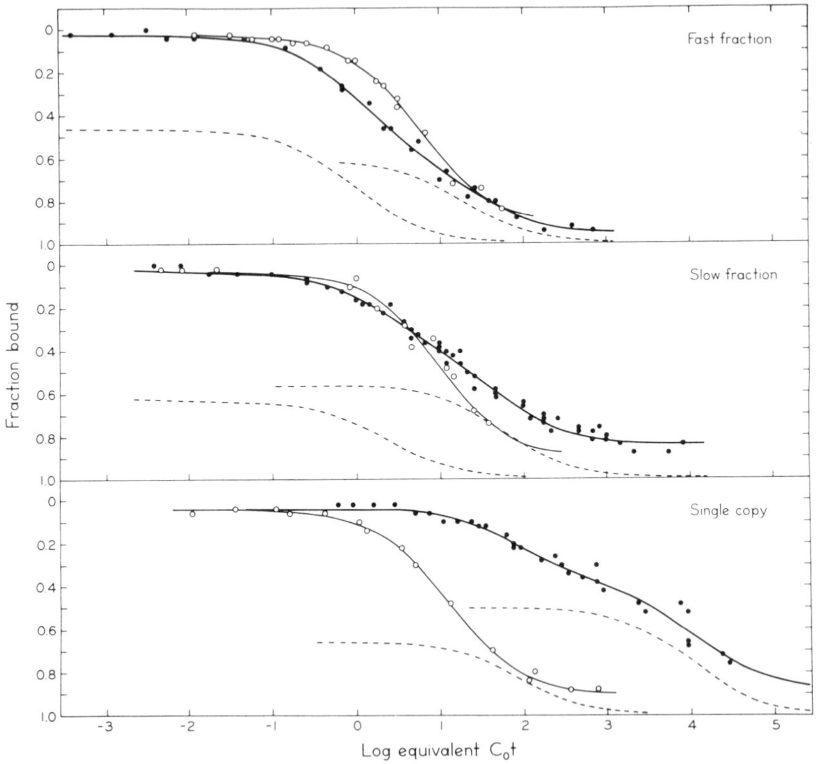

Fig. 3. Reassociation of isolated kinetic fractions of the pea genome. Short fragments of pea DNA similar to those used in the experiment of Fig. 2 were separated by hydroxyapatite chromatography into fractions corresponding to the three theoretical frequency classes illustrated in Fig. 2. DNA from each fraction was then reassociated as tracer in the presence of excess total DNA. Reassociation of these tracers is indicated by the closed symbols; open symbols refer to *E. coli* DNA tracers included as internal standards in each reaction. Repeat sequence frequency heterogeneity is indicated by the fact that none of the tracers reassociate with kinetics approximating those of a single second-order component, as would have been expected if the model describing the data in Fig. 2 were correct. (Murray et al., 1978).

least two quite different frequency classes (Preisler and Thompson, 1978; Murray et al., 1979), and similar experiments with fractions of the pea genome revealed even more kinetic heterogeneity. (Fig. 3 and Murray et al., 1978). In this case, the results are best interpreted as reflecting a more or less continuous distribution of repetition frequencies, a conclusion that has also been reached by Klein and his associates (1978) from their data on a sample of cloned repetitive sequences from the sea urchin genome.

Interspersion of Repetitive and Single-Copy Sequences

As early as 1965, Bolton and his associates observed that long fragments of eukaryotic DNA could form large multimolecular aggregates or networks after partial reassociation, and they concluded that the repetitive sequences responsible for base pairing under their conditions must be scattered throughout the genome. Experiments with mouse DNA fragments of several different lengths suggusted that the repetitive sequences in this genome occurred with an average spacing of about 1000 nucleotides or less. Interest in sequence organization was greatly stimulated in 1969, when Britten and Davidson proposed that repetitive sequences dispersed throughout the genome might function as gene regulatory elements. Subsequent studies have shown extensive interspersion of short (300–400 nucleotide) repetitive sequences at intervals of 2000 or fewer base pairs in the genomes of a wide variety of unrelated species (Davidson et al., 1975).

The ability of hydroxyapatite to bind reassociated molecules that are only partially double stranded provides a powerful tool for studying repetitive sequence interspersion. Since reassociation of a short repetitive element adjacent to a region of single-copy DNA will cause both sequences to bind to HAP, it is possible to score the fraction of fragments that contain repeats. Fig. 4 shows that this fraction will be a sensitive function of fragment length when the genome is characterized by extensive short-period interspersion, but it shows much less length dependence when most sequences are interspersed only at long intervals.

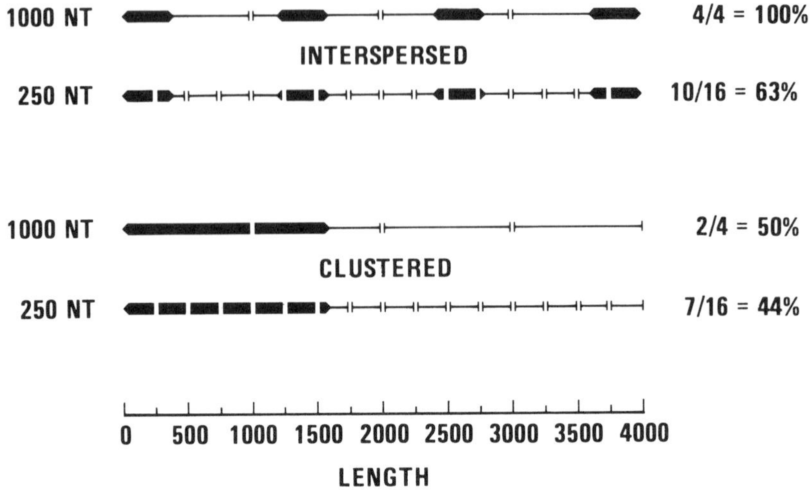

Fig. 4. Diagram to illustrate the consequences of short-period (interspersed) and long-period (clustered) patterns of sequence organization for reassociation experiments monitored by hydroxyapatite binding. Both arrangements illustrated contain 40% repetitive DNA (heavy lines). In each case, the top line represents a piece of DNA broken into fragments 1000 nucleotides long, while the lower line shows the same DNA broken into 250-NT segments. As indicated on the right, the fraction of fragments containing some portion of a repeated sequence is a much more sensitive function of fragment length in the interspersed sequence arrangement. (From Thompson and Murray, 1979; by permission.)

Fig. 5 shows a series of experiments in which we used this approach to evaluate the extent of repeat sequence interspersion in the mung bean genome. About 65% of the 300-nucleotide fragments of mung bean DNA reassociate with single-copy kinetics. The observed rate constant of 0.0016 suggests a haploid genome size of 0.5 pg, in good agreement with cytophotometric measurements (M. D. Bennett, personal communication). Since the reassociation rate of long single-copy tracer sequences increases almost directly in proportion to fragment length (Hinnebusch et al., 1978; Chamberlin et al., 1978), we would expect 1200-nucleotide fragments containing only single-copy DNA to react with a rate

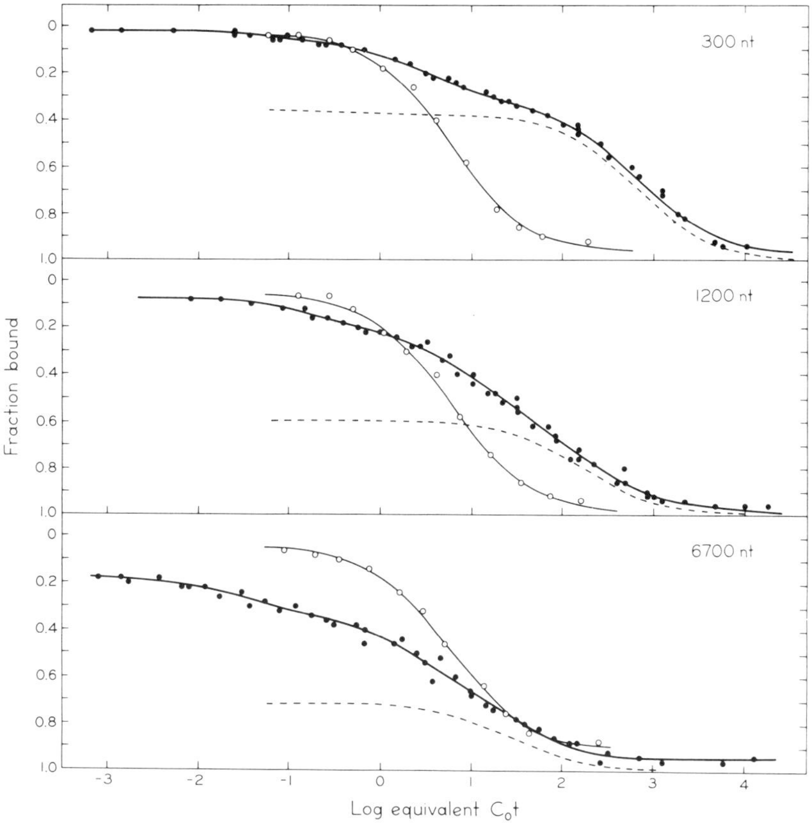

Fig. 5. Reassociation kinetics of mung bean DNA tracers of different lengths. Tracers of the indicated modal single-strand length were reassociated in the presence of excess unlabeled mung bean DNA (300 nucleotides). Knowing the single-copy rate constant for 300-nucleotide fragments, the fraction of fragments reassociating with single-copy kinetics in each of the longer tracers was estimated by fitting a component with the rate constant expected for single-copy sequences of that length. Heavy solid lines represent the best three-component solutions including this constraint, and the single-copy component is illustrated by the dashed lines. Light solid lines and open circles refer to the reassociation of *E. coli* internal standards. (Murray and Thompson, 1979.)

constant of 0.0064, and the expected rate for those 6700 nucleotides long is 0.036. Least-squares analysis of the data using components having these expected rate constants indicates that over 40% of the 1200-nucleotide fragments reassociate as predicted for single-copy sequences while some 30% still do so even when the length is increased by about five times to 6700 nucleotides. Thus, about 25%, 10%, and 30% of the mung bean genome is composed of single-copy sequences 300-1200, 1200-6700, and >6700 nucleotides long, respectively.

In contrast to the predominantly long-period interspersion pattern seen in mung bean DNA, it is possible to show that repeated sequences in pea DNA are interspersed at very short intervals throughout essentially the entire genome. Fig. 6A shows an experiment carried out with a 1300-nucleotide tracer driven by an excess of 300-nucleotide unlabeled DNA. The reassociation kinetics of the driver fragments are indicated by the heavy, solid line (taken from Fig. 2) while the line through the data points represents the best least-squares fit to the tracer data. The two theoretical components of this solution are indicated by the dashed lines. In striking contrast to the results just presented for mung bean DNA, the slowest component of this tracer reaction has a rate constant some 14 times greater than expected for 1300-nucleotide single-copy DNA, and we cannot detect the presence of *any* single-copy sequences greater than or equal to 1300 nucleotides in length.

A more sensitive assay for long single-copy sequences is shown in Fig. 6B. In this case, the tracer is composed of the most slowly reacting 18% of the fragments, which were isolated from a tracer preparation similar to that of Fig. 6A by partial reassociation and hydroxyapatite fractionation at C_0t 10. The resulting fragments, having an average length of 1000 nucleotides, were then reassociated with an excess of fresh unfractionated driver DNA. This tracer should be enriched more than five times for long single-copy sequences, and, in fact, it is possible to force the computer to include a small single-copy fraction when fitting these data with second-order components (Murray, et al., 1978). The component illustrated in the figure accounts for 6% of the tracer, or about 1% of the total DNA. However, the data can be described at least as well with only one repetitive component. Thus, even at this in-

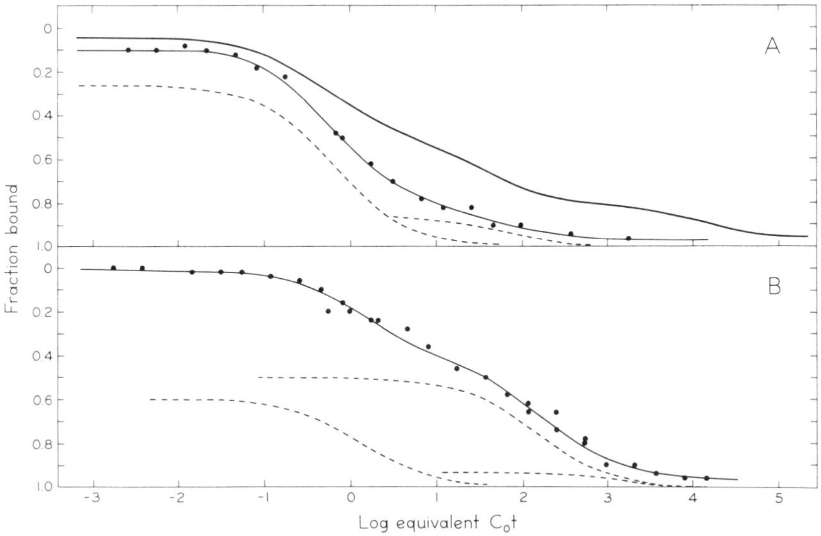

Fig. 6. Reassociation of long tracer fragments of pea DNA. Experiments similar to that of Fig. 5 were carried out with pea DNA. In each case, the reactions were standardized with *E. coli* DNA internal controls. (A) Fragments 1300 NT long were separated by alkaline agarose electroporesis and reassociated with excess 300-nucleotide driver DNA. The upper curve is reproduced from Fig. 2 to show the reassociation of short fragments. The light line is an unconstrained fit of two theoretical components to the long tracer data (●). The slowest of these components has a rate constant of about 0.01, while single-copy DNA fragments 1300 nucleotides long would reassociate with a rate of about 0.009 (see text). (B) In this experiment, the tracer was composed of fragments 1000 nucleotides long that were isolated as single strands after incubation with unlabeled driver DNA to C_0t 10. These fragments were then reassociated in the presence of an excess of fresh driver. The fit shown was obtained by forcing the computer to use a component reacting at the expected single-copy rate (see text), but equally good fits can be obtained with only repetitive components. (Murray et al., 1978.)

creased resolution, we are unable to obtain clear evidence for any single-copy sequence longer than about 1000 nucleotides in the pea genome.

Distribution of Single-Copy Sequence Lengths

When the fraction of fragments bearing repetitive elements is known for a variety of different tracer lengths, it is possible to determine the average length of single-copy sequences between repeats. The theoretical basis for this kind of analysis has been well described by Davidson and his associates (1973) and Graham and her associates (1974). When repetitive and single-copy sequences are interspersed with one another, the fraction of fragments bearing at least one repetitive element increases with increasing fragment length until the point at which the fragments begin to exceed the length of the single-copy sequences in the shortest length class. Beyond this point, all fragments containing single-copy sequences of this length also contain repetitive sequences, and the slope of the curve relating hydroxyapatite binding to fragment length shows a sharp decrease.

Thus, such binding curves provide an elegant means of analyzing the length distribution of single-copy sequences. However, it is essential to recognize that any such analysis depends on the accuracy with which the fraction of fragments bearing repetitive elements can be measured. Typically, binding curves are constructed from hydroxyapatite fractionation data obtained after incubation to a single C_0t value. The C_0t value is normally chosen to permit a reasonable separation of repetitive and single-copy DNA fragments in the short driver DNA preparation. Although it is commonly assumed that the degree of separation achieved at a particular C_0t value will remain the same for tracers of various lengths, few genomes are so conveniently designed.

Severe complications may sometimes arise from the length effects on tracer-reassociation rates mentioned above. Since long single-copy sequences, long repeats, and clusters of short repeats of similar frequency reassociate more rapidly in long tracers than they do in the short driver DNA, a C_0t value that gives an adequate separation of repetitive and single-copy fragments 300 nucleotides long may fail to do so when the fragment length is increased to 1500 or 3000 nucleotides. The mung bean data in Fig. 5 provide a good example of such a situation. Separation of repeats and single-copy sequences at C_0t 30 would be quite satisfactory for 300-

nucleotide fragments; however, a significant fraction of the 1200-nucleotide single-copy fragments would have reassociated at this C_0t and about half of the purely single copy fragments would have reacted in the longest tracer. In the absence of a full kinetic analysis, such simple binding measurements could seriously overestimate the fraction of fragments containing repetitive elements.

In some cases, such difficulties may be minimized by working at lower C_0t values. However, problems may be encountered even at lower C_0t values if the amount of repetitive DNA remaining unreacted is large in relation to the size of any of the single-copy length classes under study.

This point can be illustrated by the pea DNA data in Fig. 7, in which binding data obtained at two different C_0t values are compared. At C_0t 50, about 85% of the repetitive sequences, but only about 1% of the single-copy sequences, have reacted on 300-nucleotide fragments (Fig. 2 and Murray et al., 1978), and the expected contribution from reassociation of any long single-copy sequences would be less than 2% of the genome for fragments up to 3000 nucleotides in length. C_0t 50, therefore, establishes conditions similar to those in most published experiments of this type. If the separation of repeats and single-copy sequences were unaffected by tracer length, the slight slope of the curve above 1000 nucleotides could indicate that as much as 13% of the genome is composed of single-copy sequences with lengths in excess of 2500 nucleotides (Graham et al., 1974). Similar data have often been used to infer the presence of small fractions of long single-copy sequences in other plant genomes. However, any long regions of slow repetitive DNA (long individual sequences or clusters of short ones) would be expected to reassociate more rapidly on long tracers and thus would also contribute to the observed binding increases at the larger fragment lengths. Either or both of these effects would lead to an underestimate of the actual repetitive fraction and a simultaneous overestimate of the length dependence at longer lengths.

When the binding experiment with pea DNA is repeated at C_0t 680 the binding beyond 1000 nucleotides remains constant at 94%, which is very close to the expected maximum after a normal amount of "unreactable" DNA is allowed for (Davidson et al., 1973).

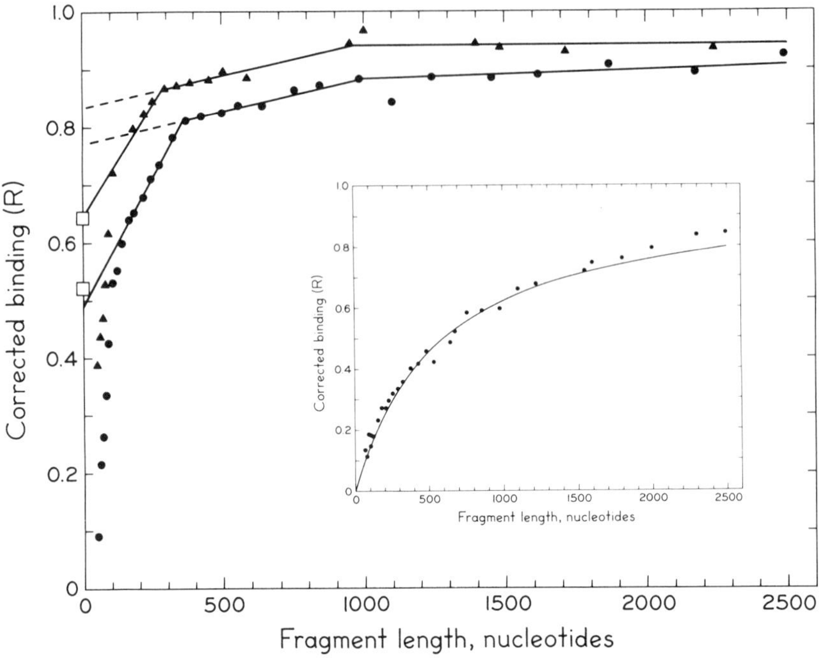

Fig. 7. Hydroxyapatite binding of reassociated pea DNA as a function of tracer fragment length. Tracers separated according to length by alkaline agarose electrophoresis were reassociated with unlabeled 300-nucleotide DNA to C_0t 50 (●) or C_0t 680 (▲). After correction for zero time binding, the fraction bound is plotted as a function of the fragment length determined after reassociation. In addition, the fraction of base-paired nucleotides was calculated from optical hyperchromicity measurements at each C_0t. These values (□) should agree with those obtained by extrapolating the binding data to zero fragment length and thus help support the conclusion that HAP binding is quantitative at lengths above about 100-150 nucleotides. The inset shows a similar experiment with *E. coli* tracers designed to illustrate length effects in the absence of repeat sequence interspersion, such as might be anticipated for incompletely reassociated components in a complex eukaryotic genome. The solid line is a theoretical binding curve predicted from the assumption that tracer reassociation rates are directly proportional to length. (Murray et al., 1978.)

Thus, the putative "long single-copy component" appears to be an artifact attributable to incomplete repetitive sequence reassociation at C_0t 50. This conclusion is supported by the long-tracer kinetic experiments described above, which directly demonstrate the absence of detectable quantities of single-copy sequences longer than 1000 nucleotides.

Both the C_0t 50 and C_0t 680, data show a sharp breakpoint at about 300-400 nucleotides. At least for fragments above about 150 nucleotides, this result cannot be explained by failure of the shorter fragments to bind to HAP. This conclusion is supported by model experiments with *E. coli* DNA fragments and by the fact that extrapolation to the Y axis from the binding data between 150 and 300 nucleotides yields an estimate of the fraction of nucleotides paired that is in excellent agreement with that obtained from independent measurements by hyperchromicity. Furthermore, the similarity of the binding curves at both low and high C_0t values and the sharpness of the break in slope (in contrast to the smooth curve obtained for *E. coli* DNA) suggest that purely kinetic effects on the reassociation of these relatively short tracers are not significant. Thus, we conclude that most of the single-copy sequences in pea DNA are very short, with an average length of about 300-400 nucleotides.

Although *repetitive* sequences with a modal length of 300-400 nucleotides are a prominent feature of plant and animal DNA (Thompson and Murray, 1979), such a short class of *single-copy* sequences is not usually reported even in highly repetitive genomes. In at least some cases, however, short single-copy elements may have been missed because of scatter in the data, the use of sheared tracers with very broad length distribution, or the absence of enough data points at very short fragment lengths. We believe that more extensive investigation might reveal classes of very short single-copy DNA in other highly repetitive plant genomes. Indeed, it is already possible to infer from the data of Smith and Flavell (1977) that a significant class of 300-400-nucleotide single-copy DNA may exist in rye.

At this point, we are confronted with an apparent paradox. Most of the single-copy DNA in the pea genome exists in the form of very short sequences, and there is no single-copy DNA detectable above about 1000 nucleotides. However, most messenger RNA in organisms is transcribed principally from single-copy sequences, and the *average* message is generally considered to be about 1200 nucleotides in length (Goldberg et al., 1973, 1978). Thus, it was important to determine whether or not mRNA is transcribed from single-copy sequences in the pea genome. This was accomplished by preparing complementary DNA copies of mRNA *in vitro* and reassociating the resulting tracer with a vast excess of genomic DNA fragments. The results (Fig. 8) show that, as in all other organisms so far examined, most of the mRNA in peas is transcribed from single-copy DNA. Since we are unable to detect single-copy sequences long enough to serve as templates for an average message, we must consider the possibility that mRNA molecules might be assembled by splicing together RNA transcripts from a number of short regions of single-copy DNA. Recent evidence (Darnell, 1978) indicates that splicing mechanisms are

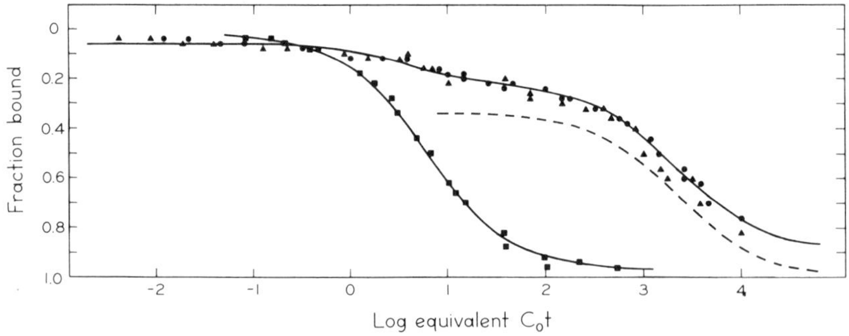

Fig. 8. Reassociation of complementary DNA synthesized from the polysomal poly A messenger RNA of pea buds. Polysomal RNA was fractionated on an oligo dT-cellulose column and the bound fraction used to prepare cDNA tracer in an oligo dT-primed reaction with reverse transcriptase from avian myeloblastosis virus. Tracer with a modal length of about 650 nucleotides as measured by alkaline agarose electrophoresis was reassociated with an approximately 3000-fold mass excess of unlabeled genomic DNA together with an internal *E. coli* kinetic standard (■). Reactions were carried out at both $T_m - 25°$ (●) and $T_m - 35°$ (▲); for simplicity, only the fit to the $T_m - 25°$ data is illustrated. This fit indicates that about 75% of the reactable tracer reassociates as expected for single-copy sequences 650 nucleotides long. The 11% unreactable fraction is probably attributable mainly to very abundant mRNA sequences since genomic DNA sequence excess would have not been achieved for polysomal RNAs present in more than about 700 copies per cell. (Peters et al., 1979.)

quite common in animal systems. However, the large size of the pea genome requires us to view such speculations with extreme caution. We cannot eliminate the possibility that 2% or 3% of the pea genome might be composed of long single-copy sequences undetectable in our experiments; even 2% would include almost as much DNA as is present in the entire *Drosophila* genome and provide enough coding capacity for 70,000 average genes.

Interspersion Patterns in Plant DNA

Table 2 presents a summary of the available data on interspersion of single-copy DNA sequences in higher plant genomes. It is immediately apparent that a significant portion of the single-copy sequences in all these genomes is contained in one or more classes with modal lengths of less than about 2000 nucleotides. Thus, the higher plant genomes studied to date correspond more closely to the "*Xenopus* pattern" of short-period interspersion than they do to the extremely long period interspersion pattern seen, for example, in *Drosophila* DNA. As defined by Davidson and his associates (1975), "*Xenopus*-pattern" organisms contain a major fraction of single-copy sequences with lengths less than about 1500 nucleotides and "*Drosophila*-pattern" species contain much longer regions, often extending for more than 10,000 nucleotides without interruption.

However, to group all higher plant genomes together in the "*Xenopus* pattern" would be an unfortunate oversimplification. Even though a component of the single-copy sequences is always seen to be arranged in a short-period interspersion pattern, there are large quantitative differences in the representation of this component. In mung bean and cotton, sequences whose lengths exceed 1200-1800 nucleotides constitute a majority of the single-copy DNA and account for 40% to 60% of the genome, while in all other species (with the possible exception of soybean), such sequences include less than half of the single-copy DNA and account for less than 10% of the genome. Conflicting results have been obtained for soybean DNA (Goldberg, 1978; Gurley et al., 1979), but the highest estimate for the fraction of single-copy sequences longer

Table 2. Summary of Data on Interspersion of Single-Copy Sequences in Higher Plant DNA[a]

Species	DNA Content ($NTP \times 10^{-9}$)	Long Tracer Kinetics: Tracer Lengths	Long Tracer Kinetics: % Single Copy	Binding Experiments: Length Class	Binding Experiments: %	Reference
Mung bean	0.5	300	65			Fig. 5
		1200	42			
		6700	31			
Cotton	0.8			1800	25	Walbot and
				≥ 3600	40	Dure (1976)
Soybean	1.1	300	23	1200	40	Goldberg (1978)
		2700	0	9000	11	Gurley et al.
						1979
Parsley	1.9	300	12			Kiper and
		1350	4	1000	20	Herzfeld
		5300	0	≥ 5000	11	(1978)
Tobacco	2.0	300	22			Zimmerman
		950	19	1400	26	and Goldberg
		2400	2			(1977)
		4300	1	≥ 4300	11	
Pea	4.8	300	17	300	20	Murray et al.
		1000	≥ 3	1000	10	(1978)
		1300	0	≥ 2300	6	
Wheat	5.6			1000	15	Flavell and
				≥ 9000	10	Smith (1976)
Rye	8.6			300	13	Smith and
				1500	7	Flavell
				3500	5	(1977)
				≥ 6000	5	

[a] For long tracer kinetic experiments, data refer to the fraction of fragments of the indicated length that reassociate with single-copy kinetics. Binding data refer to the fraction of the genome composed of single-copy sequences in each length class as estimated by the extrapolation procedure of Graham et al. (1974). In the case of binding data, we have attempted to apply a uniform interpretation to data from different laboratories. Thus, the values given in the table may differ in some respects from those in the original publications.

than about 1200 nucleotides is less than 20%, which would account for less than one-third of the single-copy DNA. Thus, the small sample of plant genomes analyzed to date may be grouped into two categories, depending on whether or not a majority of the single-copy sequences are longer than about 1500 nucleotides. In fact, when we look only at complete kinetic analyses—which are probably more reliable than binding experiments—it is tempting to push the distinction even farther. Except for the mung bean, none of the five genomes analyzed in this way seem to have significant fractions of gene-length single-copy DNA.

Sequence Organization and Speculations on Genome Evolution

Fig. 9 shows a comparison of the pea and mung bean genomes in terms of the absolute amount of DNA contained in the major

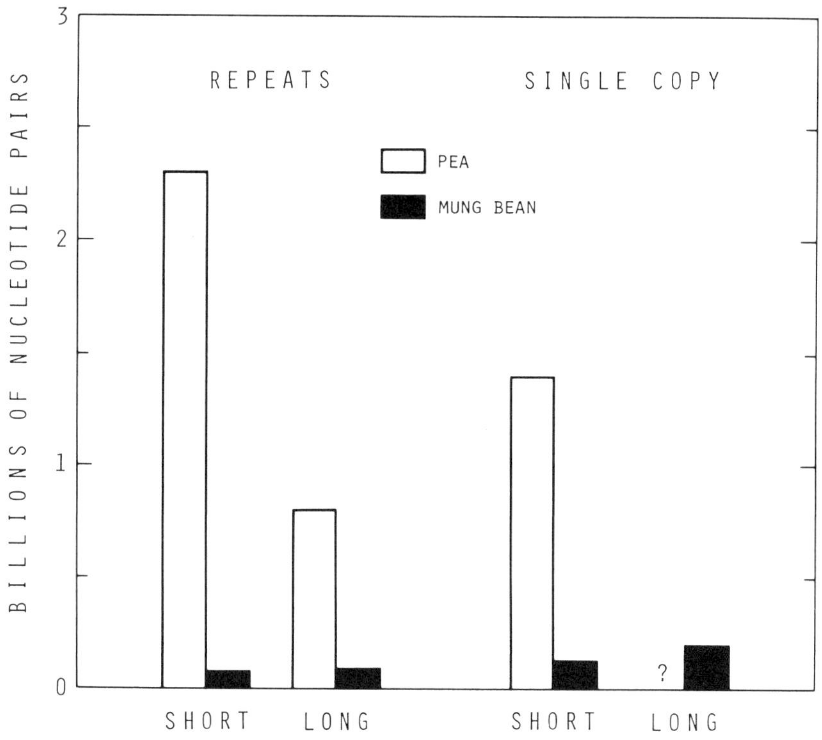

Fig. 9. Comparison of pea and mung bean genomes. The bars indicate the amount of DNA in each genome that is composed of repetitive or single-copy sequences shorter or longer than about 1200 nucleotide pairs.

repetitive and single-copy sequence classes. The nine-fold difference in total genome size means that there are some 4 *billion* more nucleotide pairs of DNA in the pea genome than in that of mung beans. It is very hard to believe that these 4 billion nucleotide pairs are all required for functions unique to pea plants. In addition, it is clear from the figure that essentially all of this extra DNA is either repetitive or composed of very short single-copy sequences. Since the majority of the repetitive sequences are also short, it is possible to imagine that most of the short single-copy sequences originated from ancient repetitive-sequence families simply by accumulating base substitutions until all experimentally recognizable homology between family members was lost. In this case, most of the single-copy sequences in the pea genome would actually be "fossil repeats."

Since this hypothesis implies that repetitive sequences undergo gradual evolutionary divergence after their original incorporation into the genome, it predicts that a broad distribution of pairing precision should be found among the repetitive duplexes formed by random reassociation *in vitro*. In fact, it is well known that such duplexes typically do melt over a rather broad range of temperatures. An example is illustrated in Fig. 10, which also indicates that these broad distributions can be resolved into theoretical components with different thermal stabilities in much the same way that theoretical frequency classes can be resolved from reassociation kinetic data (Cuellar et al., 1978). As a working hypothesis, we suppose that the thermal components seen in experiments such as that in Fig. 10 may reflect the presence of groups of repetitive-sequence families characterized by similar amounts of sequence divergence. According to the gradual-divergence hypothesis, each such group would be considered to have been produced by amplification events occurring at about the same time so that the overall profile would contain a record of the major amplification events in the history of the species. A variety of evidence in the literature suggests that amplification—and even reamplification—events do occur fairly frequently in evolution (Thompson and Murray, 1979). For example, it is usually possible to distinguish even quite closely related species on the basis of their content of different groups of repetitive sequences (e.g., Gillespie, 1977; Rimpau et al., 1978).

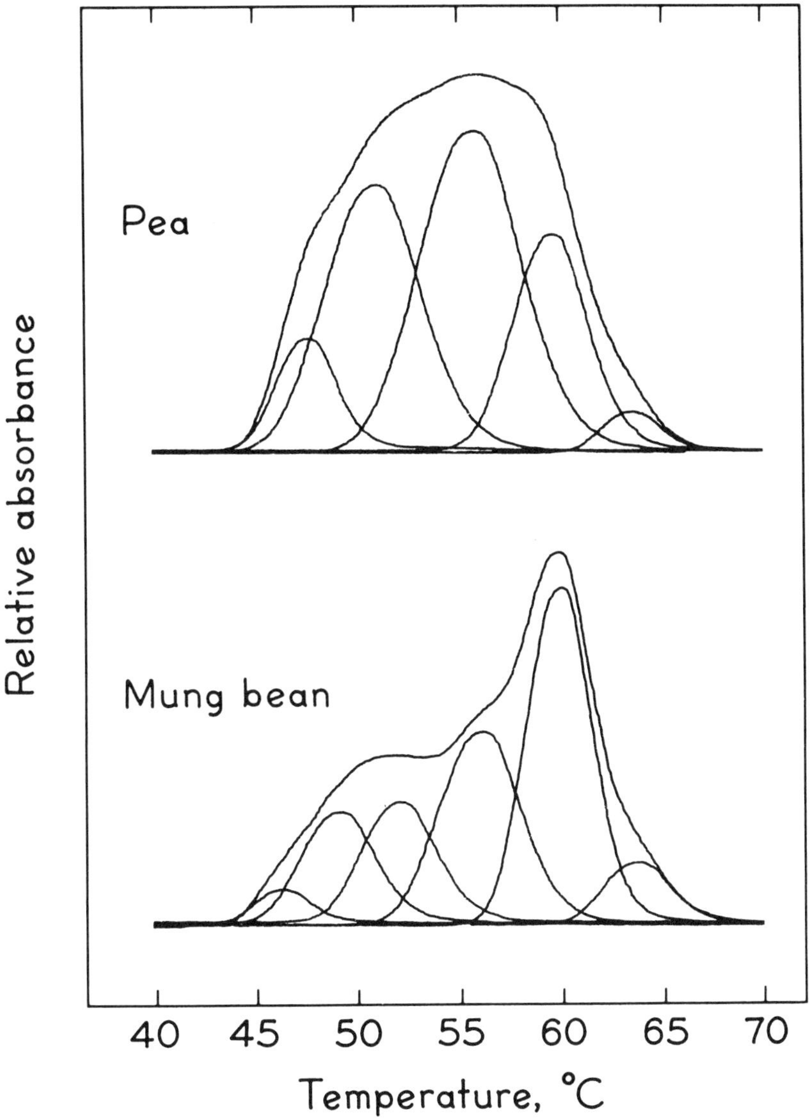

Fig. 10. First derivatives of optical thermal denaturation profiles for reassociated repetitive sequences from pea and mung bean DNA. Sheared DNA fragments were reassociated and melted in 2.4 *M* tetraethylammonium chloride to eliminate the contribution of base-composition heterogeneity to the width of the melting profile, and the results were analyzed by higher derivative procedures to provide a basis for fitting the theoretical Gaussian components shown. The techniques used have been fully described by Cuellar et al. (1978).

In addition to amplification and base substitution, genome evolution may also involve deletion events. Repeated sequences arranged tandemly in large arrays—as in many satellite DNAs, for example—may be especially subject to both amplification and deletion by means of unequal crossing-over (Smith, 1976). More generally, any sequences clustered together on a single chromosome and not subject to positive natural selection might be expected to undergo occasional deletions. Although similar random deletion events would not be likely to simultaneously affect all the members of a widely scattered repeat family, such events may well be detectable at the single-copy-sequence level.

Evidence for deletion of single-copy sequences has been obtained from a recent study of several species in the genus *Atriplex*. Fig. 11 shows some of the data on which this conclusion is based. Interspecific reactivity was assessed by measuring the complete kinetics of the reactions between single-copy tracers from various species and total DNA from either of two reference species. The kinetic data was then analyzed in the computer, using an expression designed to take into account the retardation of duplex formation that occurs in interspecific reactions involving mismatched sequences (Galau et al., 1976; Belford and Thompson, 1977). For these reactions, retardation effects were minimal, and, even after they were taken into account, it was clear that many of the interspecific reactions are far from complete. We believe the simplest explanation for these data is that deletion events occurred rather frequently during the period of evolution after speciation, with different sequences or regions of the genome being deleted in different lineages. This hypothesis is consistent with the widely accepted notion that evolutionary specialization is often associated with a decrease in total genome size (e.g., Hinegardner, 1976).

The examples just cited constitute only a small part of a rather large body of evidence, accumulated in several laboratories over the last 10 years or so, that supports the idea that rather large scale amplification, interspersion, and deletion events are prominent features of genome evolution in higher plants and many other eukaryotes (Thompson and Murray, 1979). At the same time, the tremendous variation of DNA content in different organisms supports the view that most of the DNA in a "typical" eukaryote

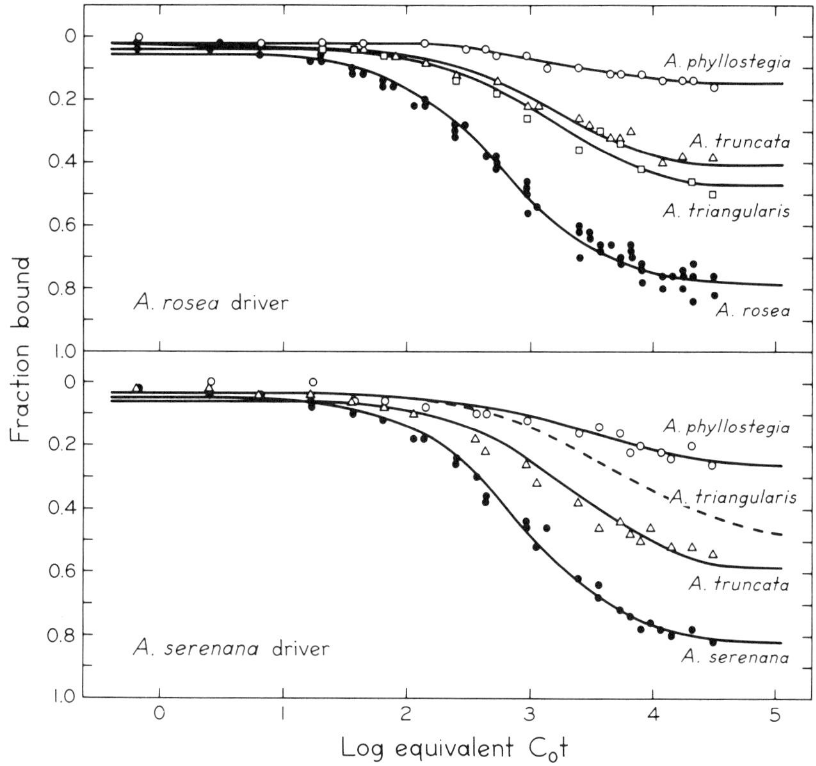

Fig. 11. Interspecific cross reactivity of single-copy DNA sequences among several species of the genus *Atriplex*. Tracers prepared from single-copy DNA of the indicated species were reassociated with an excess of unlabeled driver DNA from one of two reference species. In other experiments, the extent of base-pair mismatch in the interspecific hybrids was quantitated by measuring their thermal stability. This information was used to adjust for the reduction in tracer reassociation rate attributable to mismatch in fitting the theoretical curves shown (see text). (Belford and Thompson, 1979.)

is not necessary for basic coding and regulatory functions, and comparisons such as that between peas and mung beans suggest that even the general pattern of sequence organization may not be related in any direct way to organismic complexity or developmental programming. Such considerations have led us to emphasize evolutionary perspectives and to think largely in terms of genome evolution.

There is already enough evidence to suggest some impressive regularities in genome evolution. For example, Fig. 12 presents the relationship between total genome size and the amount of single-copy DNA for most animals and all plants for which we have com-

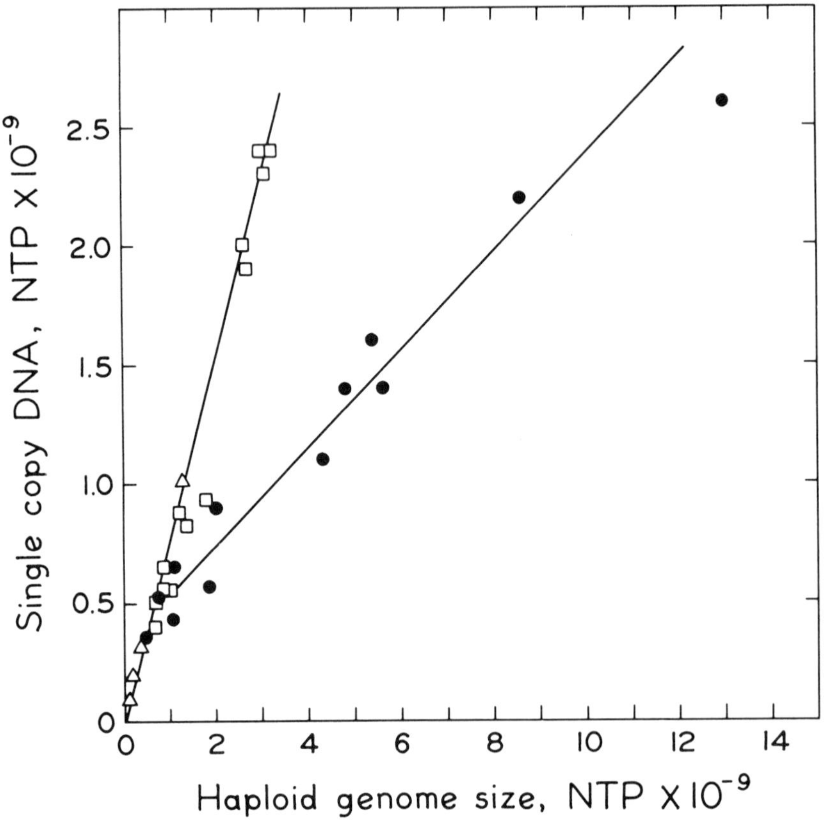

Fig. 12. Single-copy DNA content as a function of genome size in plants (filled symbols) and animals (open symbols). Animals having short-period interspersion are indicated by squares; those with long-period patterns by triangles. All values of single-copy DNA content are based on measurements that are not strongly influenced by different patterns of sequence interspersion. Data for plants were obtained from the studies references in Table 2 and the majority of the animal DNA data was taken from Davidson et al. (1975). For more information, see Thompson (1978).

plete data. The apparent consistency of the relationship is quite impressive, as is the clear difference between plants and animals that becomes apparent at genome sizes above 10^9 nucleotide pairs. If we accept the hypothesis that amplification occurs repeatedly during evolution and that single-copy sequences arise by gradual divergence of ancient repeated sequences, the different slopes of the curves for plants and animals may reflect differences in the average frequency of amplification events between the two kingdoms. Although highly speculative, this notion would help to explain the fact that plant genomes as a group tend to be larger and more variable in size than those of animals. It also receives support from the demonstration by Flavell and his associates (1977) that frequent amplifications and reamplifications have occurred during the evolution of cereal grain genomes.

At least in large genomes such as that of pea plant, it seems likely that the bulk of the DNA has no strongly sequence-dependent function and corresponds to what Hinegardner (1976) has called "secondary DNA"—that is, DNA that plays little or no direct role in the immediate activities of the organism and is free to evolve rapidly. A particularly intriguing consequence of this idea is that secondary DNA may contribute to the evolution of new organisms by providing a source of novel genetic information and/or by facilitating the rearrangement and reprogramming of existing information.

If these ideas are correct, it follows that the organization and function of the eukaryotic genome must be studied and understood in an evolutionary as well as a developmental context. In our view, complex patterns of sequence repetition and arrangement probably arise largely from the stochastic processes of molecular evolution rather than because they are essential for genotypic functions. This does not mean that functions such as gene regulation may not evolve to take advantage of these sequence patterns, and it is still possible that we may find an intimate relationship between genome organization and function. However, many more careful comparative studies will be required before we are in a position to understand fully the evolutionary component of such a relationship.

References

Belford, H. S., and Thompson, W. F. (1977) Single copy DNA sequence comparisons in *Atriplex. Carnegie Inst. Wash. Year Book 76*, 246-52.

Belford, H. S., and Thompson, W. F. (1979) Single copy DNA homologies and the phylogeny of *Atriplex. Carnegie Inst. Wash. Year Book 78* (in press).

Bolton, E. T.; Britten, R. J.; Cowie, D. B.; Roberts, R. B.; Szafranski, P.; and Waring, M. J. (1965) Renaturation of the DNA of higher organisms. *Carnegie Inst. Wash. Year Book 64*, 316-33.

Bonner, T. F.; Brenner, D. J.; Neufeld, B. R.; and Britten, R. J. (1973) Reduction in the rate of DNA reassociation by sequence divergence. *J. Mol. Biol. 81*, 123-35.

Britten, R. J., and Davidson, E. H. (1969) Gene regulation for higher cells: A theory. *Science 165*, 349-57.

Britten, R. J.; Graham, D. E.; and Neufeld, B. R. (1974) Analysis of repeating DNA sequences by reassociation. *Meth. Enzymol. 29*, 363-418.

Britten, R. J., and Kohne, D. E. (1968) Repeated sequences in DNA. *Science 161*, 529-40.

Chamberlin, M. E.; Galau, G. A.; Britten, R. J.; and Davidson, E. H. (1978) Studies on nucleic acid reassociation kinetics: V. Effects on disparity in tracer and driver fragment lengths. *Nucleic Acids Res. 5*, 2073-94.

Cuellar, R. E.; Ford, G. A.; Briggs, W. R.; and Thompson, W. F. (1978) Application of higher derivative techniques to analysis of higher resolution thermal denaturation profiles of reassociated repetitive DNA. *Proc. Nat. Acad. Sci., U.S.A. 75*, 6026-30.

Darnell, J. E. (1978) Implications of RNA·RNA splicing in evolution of eukaryotic cells. *Science 202*, 1257-60.

Davidson, E. H.; Hough, B. R.; Amenson, C. S.; and Britten, R. J. (1973) General interspersion of repetitive with non-repetitive sequence elements in the DNA of *Xenopus. J. Mol. Biol. 77*, 1-23.

Davidson, E. H.; Galau, G. A.; Angerer, R. C.; and Britten, R. J. (1975) Comparative aspects of DNA organization in metazoa. *Chromosoma 51*, 253-59.

Flavell, R. B.; Rimpau, J.; and Smith, D. B. (1977) Repeated sequence DNA relationships in four cereal genomes. *Chromosoma 63*, 205-22.

Flavell, R. B., and Smith, D. B. (1976) Nucleotide sequence organization in the wheat genome. *Heredity 37*, 231-52.

Galau, G. A.; Chamberlin, M. E.; Hough, B. R.; Britten, R. J.; and Davidson, E. H. (1976) Evolution of repetitive and nonrepetitive DNA. In *Molecular Evolution* (F. J. Ayala, ed.), pp. 200-24. Sinauer Associates, Sunderland, Massachusetts.

Gillispie, D. (1977) Newly evolved repeated DNA sequences in primates. *Science 196*, 889-91.

Goldberg, R. B. (1978) DNA sequence organization in the soybean plant. *Biochem. Genet. 16*, 45-68.

Goldberg, R. B.; Galau, G. A.; Britten, R. J.; and Davidson, E. H. (1973) Nonrepetitive sequence representation in sea urchin embryo messenger RNA. *Proc. Nat. Acad. Sci., U.S.A. 70*, 3516-20.

Goldberg, R. B.; Hoschek, G.; Kamalay, J. C.; and Timberlake, W. E. (1978) Sequence complexity of nuclear and polysomal RNA in leaves of the tobacco plant. *Cell 14*, 123-31.

Graham, D. E.; Neufeld, B. R.; Davidson, E. H.; and Britten, R. J. (1974) Interspersion of repetitive and nonrepetitive DNA sequences in the sea urchin genome. *Cell 1* 127-37.

Gurley, W. B.; Hepburn, A. G.; and Key, J. L. (1979) Sequence organization in the soybean genome. *Biochem. Biophys. Acta* (in press).

Hinegardner, R. (1976) Evolution of genome size. In *Molecular Evolution* (F. J. Ayala, ed.), pp. 179-99. Sinauer Associates, Sunderland, Massachusetts.

Hinnebusch, A. G.; Clark, V. E.; and Klotz, L. C. (1978) Length dependence in reassociation kinetics of radioactive tracer DNA. *Biochemistry 17*, 1521-29.

Kiper, M., and Herzfeld, F. (1978) DNA sequence organization in the genome of *Petroselinum sativum* (Umbrelliferae). *Chromosoma 65* 335-51.

Klein, W. H.; Thomas, T. L.; Lai, C.; Scheller, R. H.; Britten, R. J.; and Davidson, E. H. (1978) Characteristics of individual repetitive sequence families in the sea urchin genome studied with cloned repeats. *Cell 14*, 889-900.

Marsh, J. L., and McCarthy, B. J. (1974) Effect of reaction conditions on the reassociation of divergent deoxyribonucleic acid sequences. *Biochemistry 13*, 3382-88.

Murray, M. G.; Cuellar, R. E.; and Thompson, W. F. (1978) DNA sequence organization in the pea genome. *Biochemistry 17*, 5781-90.

Murray, M. G., and Thompson, W. F. (1979) DNA sequence organization in the mung bean genome. *Carnegie Inst. Wash. Year Book 78* (in press).

Murray, M. G.; Palmer, J. D.; Cuellar, R. E.; and Thompson, W. F. (1979) Deoxyribonucleic acid sequence organization in the mung bean genome. *Biochemistry 18*, 5259-66.

Pearson, W. R.; Davidson, E. H.; and Britten, R. J. (1977) A program for least squares analysis of reassociation and hybridization data. *Nucleic Acids Res. 4*, 1727-35.

Peters, D. L.; Murray, M. G.; and Thompson, W. F. (1979) The use of cDNA for studies of transcribed sequences. *Carnegie Inst. Wash. Year Book 78* (in press).

Preisler, R. S., and Thompson, W. F. (1978) Distribution of nucleotide sequence divergence among families of repetitive sequences in mung bean DNA. *Carnegie Inst. Wash. Year Book 77*, 323-30.

Rimpau, J.; Smith, D.; and Flavell, R. (1978) Sequence organization analysis of the wheat and rye genomes by interspecies DNA/DNA hybridization. *J. Mol. Biol. 123*, 327-59.

Smith, D. B., and Flavell, R. B. (1977) Nucleotide sequence organization in the rye genome. *Biochem. Biophys. Acta 474*, 82-97.

Smith, G. P. (1976) Evolution of repeated DNA sequences by unequal crossover. *Science 191*, 528-35.

Sparrow, A. H.; Price, H. J.; and Underbrink, A. G. (1972) A survey of DNA content per cell and per chromosome of prokaryotic and eukaryotic organisms: Some evolutionary considerations. *Brookhaven Symp. Biol. 23*, 451-93.

Thompson, W. F. (1978) Perspectives on the evolution of plant DNA. *Carnegie Inst. Wash. Year Book 77*, 310-16.

Thompson, W. F., and Murray, M. G. (1979) The nuclear genome: Structure and function. In *The Biochemistry of Plants,* vol. 6 (A. Marcus, ed.). Academic Press, New York (in press).

Walbot, V., and Dure, L. S. (1976) Developmental biochemistry of cotton seed embryogenesis and germination: VII, Characterization of the cotton genome. *J. Mol. Biol. 101*, 503-36.

Wilson, D. A. and Thomas, C. A. (1973) Hydroxyapatite chromatography of short double-helical DNA. *Biochem. Biophys. Acta 331*, 333-40.

Zimmerman, J. L., and Goldberg, R. B. (1977) DNA sequence organization in the genome of *Nicotiana tabacum. Chromosoma 59*, 227-52.

Cereal Genome Studies and Plant Breeding Research

R. B. Flavell, M. O'Dell, and J. Jones

There are few pieces of fundamental research that have directly benefited the production of economically successful crop plant varieties. There are many reasons for this. The biological processes on which plant breeding and crop performance depend are complex, and we understand relatively little about them at the molecular, cellular, or whole-plant level. With such an inadequate knowledge of these processes, it is difficult to design specific experiments that have incisive repercussions for plant breeding. Another reason, which surely should not continue to be ignored, is that very little basic plant research has been directly inspired by specific plant breeding problems or carried out alongside plant breeding programs. It seems inevitable that, unless more fundamental plant research is carried out closely aligned with plant breeding research, the chances of plant breeding gaining direct help from fundamental endeavors will remain limited.

At the Plant Breeding Institute in Cambridge, many basic research

Note: We are grateful to numerous colleagues past and present who have contributed to the development of this work. We thank John Bedbrook for the molecular cloning of the rye repeated sequences and for critically reading our manuscript.

projects are being actively pursued, but almost all are on plant species for which there are ongoing plant breeding programs for commercial variety production. Frequently, the fundamental and applied aspects are bridged by genetic and physiological programs that focus on characters of agronomic importance. This is especially so in the institute's wheat programs.

One of the aims of the wheat genetic studies for plant breeding has been the widening of the gene pool available to breeders by incorporating genes from distantly related accessions or species. The abilities of hexaploid bread wheat to tolerate aneuploidy and to hybridize with other related species have enabled cytogeneticists to incorporate alien chromosomes or parts of alien chromosomes into wheat (Riley and Kimber, 1966). For example, wheat lines possessing whole chromosomes, chromosome arms, or segments of chromosomes from rye (Riley and Chapman, 1958; 1960; Driscoll and Jensen, 1963; Driscoll and Sears, 1971), *Aegilops umbellulata* (Sears, 1956), *Aegilops comosa* (Riley, Chapman, and Johnson, 1968), and *Agropyron elongatum* (Knott, 1961; Sears, 1972; Dvojak and Knott, 1974), have been created and explored to some extent for their value in wheat breeding.

The incorporation of foreign chromosome material into wheat has been detected by (1) changes in the wheat karyotype, (2) the pairing during meiosis of the alien chromosomes with marker chromosomes in tester stocks, (3) the presence of new pieces of chromatin recognizable by Giemsa staining techniques (Darvey and Gustafson, 1975; Singh and Robbelen, 1976), (4) the presence of new isozyme or protein variants (for example, Bhatia and Smith, 1966; Hart, McMillin and Sears, 1976), and (5) the presence of other new genetic traits known to be present in the donor species. With the exception of methods 4 and 5, these methods are capable of detecting only relatively large pieces of chromosome and method 2 takes a long time. The incorporation of large segments of alien genomes into wheat is usually detrimental for the wheat plant. For plant breeding purposes, the incorporation of much smaller pieces of DNA containing small numbers of desirable genes is a much more attractive goal. Where such genes are easily detected by a phenotypic change, there is no difficulty in inferring the presence of the alien DNA. However, in the absence of such marker genes, alternative sensitive approaches are required to detect relatively small amounts of alien DNA.

During the last few years, many new biochemical techniques have been developed, which make detection of certain types of alien DNA in plant genomes relatively straightforward. The simplest approaches based on these techniques use DNA sequences from the donor species, absent from the recipient species, to "probe" for the presence of donor DNA in the recipient species. The success of this approach depends upon having knowledge of the donor DNA sequences that form the probes. In this paper, we describe the use of repeated DNA sequences as probes for detecting alien DNA in wheat. Repeated sequence probes are easier to use than nonrepeated or gene sequence probes because detection of a family having hundreds or thousands of essentially identical copies is technically considerably easier than detection of a single copy.

Species-Specific Repeats in *Triticinae* Species

Different *Triticinae* species can be distinguished from one another by analysis of the families of repeated DNA sequences (Flavell, Rimpau, and Smith, 1977). Under the hybridization conditions commonly used to study DNA renaturation, over 70% of the DNA in cereal genomes consists of reiterated sequences in excess of 500 copies per haploid genome. The number of families of repeated sequences shared by two species is related to the phylogenetic distance between the species; that is, more closely related species have more repeated sequence DNA families in common (Flavell, Rimpau, and Smith, 1977). In this paper, we focus on the families of repeated sequences that are species-specific, that is, not found in the other species in question.

The proportions of species-specific repeated DNA in wheat, rye, barley, and oats measured under defined hybridization conditions are shown in Table 1. These repeated sequences are found (1) clustered and not interspersed with repeats found in other species and (2) interspersed at short intervals with repeated sequences that are also found in other species (see Table 1). This implies that the proportion of the genome marked by the species-specific repeats is considerably greater than that occupied by the repeats themselves. Estimates of these proportions are given in Table 2. They are compiled from the species-specific repeat data shown in Table 1 and the proportions of other sequences with which the species-specific repeats are finely interspersed in the chromosomes. These latter values were determined by Rimpau, Smith, and Flavell (1978) for wheat and rye and in experiments as yet unpublished for barley and oats. Each value in Table 2 can be considered to be the fraction of DNA fragments that contain species-specific repeated sequences if each of the genomes were sheared into fragments a few thousand base pairs long (10^4 base pairs = 0.01% of the barley genome). Hence, if random oats or barley DNA fragments a few thousand base pairs long were incorporated into a wheat nucleus, there would be around a 75% and 40% chance, respectively, that they would carry species-specific repeats. We have carried out simple reconstruction experiments, mixing different amounts of barley DNA with wheat DNA and using total barley repeated sequences to "probe" for the barley DNA. As little as 0.05% barley DNA can be routinely detected (Flavell et al., 1978). No doubt the sensitivity could be increased with modifications to the hybridization conditions and with more-specific probes. These sorts of ex-

Table 1. Proportions of Genomes Consisting of Species-Specific Repeats

	Interspersed[b]	Clustered[c]	Total[a]
Wheat	9	7	16
Rye	16	7	23
Barley	22	6	28
Oats	37	18	55

[a] The proportions of the genomes consisting of species-specific repeats were determined from interspecies DNA/DNA hybridization experiments. The hybridizations were performed at 60 °C in 0.12 *M* phosphate buffer to C_0t values, which allow virtually all fragments containing a repeated sequence to renature. The proportions of renatured DNA were determined by hydroxyapatite chromotography and corrected for the effects of DNA fragments length (see Flavell, Rimpau, and Smith, 1977; Rimpau, Smith, and Flavell, 1978).

[b] These repeats are interspersed with relatively short repeats found in another species or with nonrepeats (see Rimpau, Smith, and Flavell, 1978).

[c] These repeats are considered clustered because they are not closely interspersed with repeats found in another species (see Rimpau, Smith, and Flavell, 1978).

Table 2. Estimates of the Proportions of *Triticinae* Genomes in which Species-Specific Repeats are Concentrated[a]

Wheat	28
Rye	34
Barley	41
Oats	74

[a] The estimates are the sums of the total figures in Table 1 and the amounts of short repeated and nonrepeated sequences with which some of the species-specific repeats are interspersed (see text and Rimpau, Smith, and Flavell, 1978 for estimates of these amounts).

periments could provide definitive proof for the incorporation of barley DNA into the wheat genome. Such proof will be of use to laboratories specifically interested in constructing wheat-barley hybrids and wheat lines containing disomic additions of barley chromosomes (Islam, Shepherd, and Sparrow, 1975; Chapman and Miller, 1978). The sensitivity of the DNA/DNA hybridization methods is clearly several orders of magnitude more than of any cytological method.

The Distribution of Rye-Specific Repeats in the Rye Karyotype

To assess further the value of species-specific repeated sequences as probes for alien DNA, knowledge of the distribution of the repeated sequences in the alien genome is important. This is particularly relevant where the proportion of species-specific repeats in the genome is relatively small, for example, in rye and wheat. Therefore, we have investigated the distribution of rye-specific repeats by exploiting wheat lines carrying single homologous pairs of rye chromosomes. When sheared tritium labeled rye DNA is incubated with large excesses of sheared, unlabeled DNA from wheat and each rye addition line, the proportion of rye repeated sequences hybridizing to the unlabeled DNAs is significantly greater for any line possessing a rye chromosome than for wheat alone. The value was $64.9 \mp 1.5\%$ when unlabeled wheat DNA was used but 68.9% to 75.2% when a single rye chromosome was also present (Flavell et al., 1978). These latter values were all significantly higher than the 64.9% wheat DNA value, but the difference is rather small.

A large fraction of the rye-specific repeated sequences is present in very high copy numbers in the rye genome. Thus, the detection sensitivity of rye-specific sequences in wheat can be enhanced by enriching for the highly repetitive DNA. A fraction of the rye genome enriched for these sequences can be isolated easily by renaturing sheared DNA to a low C_0t value (e.g., 10^{-2}) and isolating the reannealed highly repetitive DNA by hydroxyapatite chromatography (Flavell et al., 1978). When this DNA is used to detect the presence of rye DNA in wheat-rye addition lines, quantitatively larger differences are seen. In Table 3, results are presented from an experiment in which the unlabeled DNAs were bound to nitrocellulose filters and incubated together with rye C_0t 10^{-2} DNA, radioactively labeled *in vitro* by nick translation. Fairly stringent hybridization conditions were used (2 × SSC 50% formamide at 35 °C) to ensure hybridization of only very closely related nucleotide sequences. A 100-fold excess of sheared unlabeled wheat DNA was mixed with the radioactive rye DNA in the incubation solution to reduce the amount of labeled rye DNA binding to sequences in the wheat genome. The results clearly show that the presence of a single arm of any rye chromosome in a wheat genome background can easily be detected. Thus, rye-specific repeated sequences are present in all rye chromosome arms.

Rye chromosomes have relatively large blocks of telomeric-heterochromatin (Bennett, Gustafson, and Smith, 1977) that consist largely of highly repeated sequences. About 15% of the DNA in the genome is localized in this heterochromatin (Bennett, Gustafson, and Smith, 1977), so there is a very high probability that rye-specific highly repeated sequences are localized in these chromosomal regions. Thus, highly repeated sequence rye probes are likely to detect mainly rye heterochromatin.

In our laboratory, two major families of rye-specific repeated sequences have been studied in detail (Bedbrook et al., 1980). The first contains over half a million copies of a unit sequence 480 nucleotide pairs long. It accounts for about 5% of the total DNA of the rye genome. The second occupies about 2% of the rye genome, and the unit sequence length is 615 nucleotide pairs. Analyses

Table 3. Detection of Single Rye Chromosomes in a Wheat Genetic Background Using Highly Repeated Rye DNA as a Probe

Rye Chromosome Added to Wheat	Level of Hybridization[a]	Rye Chromosome Added to Wheat	Level of Hybridization
–	100	5	170*
1	223*	5^L	143*
1^L	180	5^S	157
2	144*	6	201*
2 right	213*	6^L	146*
2 left	200*	6^S	173
3	192*	7	187*
3^S	137	7^L	183
4	145*		
4^L	188*		
4^S	149*		

[a] Unlabeled denatured DNA from each of the wheat-rye addition lines was loaded on to five replicate nitrocellulose filters. These were incubated altogether with 0.37 μg/ml of denatured highly repeated rye DNA that had been radioactively labeled *in vitro* by nick translation. Incubation was carried out in 50% formamide, 2 × SSC at 35 °C for 18 hours. The radioactive DNA bound to each filter per μg of unlabeled DNA was calculated and is expressed in the table relative to the mean amount of rye DNA hybridized per μg of filter-bound wheat DNA. All hybridization levels of rye DNA to unlabeled addition line DNAs were significantly greater than to wheat DNA at the 1% level of probability except those marked *, which were significantly greater at the 0.1% level of probability. For further details of this experiment, see Flavell et al., 1978. L = long arm of the rye chromosome. S = short arm.

of the organization of these sequences in the rye chromosomes has been greatly facilitated by the molecular cloning of members of these two families in *Escherichia coli*, using the cloning vector plasmid pBR322. Some of the details of the procedures used to clone these repeats are included elsewhere in this volume (Bedbrook et al., 1980). The localization of these repeats in the chromosomes has been determined by *in situ* hybridization. Highly radioactively labeled RNA copies of the bacterial plasmids containing the rye repeats were prepared using *E. coli* RNA polymerase. These RNAs were then incubated with denatured metaphase chromosomes, immobilized on glass slides. The sites of hybridization of the RNAs were recognized by autoradiography. Both families of rye-specific sequences are virtually confined to the heterochromatic telomeres. A photograph showing the localization of the 480 nucleotide repeats is included in Fig. 1, together with the *in situ* hybridization of the same repeat to a wheat line carrying 42 wheat chromosomes and a homologous pair of rye chromosomes. This second photograph clearly demonstrates the use of the cloned rye-specific DNA probes for distinguishing a single pair of rye chromosomes in a wheat nucleus.

The use of these pure rye-specific radioactive DNA probes in DNA/DNA hybridization reactions with either purified DNA or *in situ* in the chromosomes makes the detection of the telomeric regions of rye chromosomes straightforward. Whether other rye-specific probes can be isolated that will conveniently mark other regions of the chromosomes remains to be explored. The distribution of silver grains over the rye chromosomes in *in situ* hybridization experiments is different for different chromosomes, even using the highly repeated telomeric sequences as probes (Fig. 1A). It is likely, therefore, that with several probes each chromosome or chromosome arm will be able to be unambiguously differentiated from all the other chromosome arms in the karyotype (Appels et al., 1978).

The chromosomes carrying the major sites of the ribosomal RNAs and the 5S RNAs are easily characterised by *in situ* hybridization, using a bacterial plasmid containing the genes for these RNAs (Bedbrook et al., 1980, and W. L. Gerlach et al., 1978). Using repeated sequences fractionated from the wheat genome by centrifugation in Ag^+ $CsSO_4$, Gerlach and his associates (1978) have shown that the chromosomes of the B genome of hexaploid wheat are preferentially labeled in *in situ* hybridization experiments. Thus, in the not too distant future, we can expect chromosome identification by specific probe DNA hybridization techniques to become a routine part of cytogenetic analysis and also of plant breeding research in which chromosome identification is a step in a plant breeding program. Where specific chromosome identification is not required but simply the detection of alien DNA, sequences distributed over many chromosomal regions would form better probes. It is to be hoped that collections of different sorts

Fig. 1A. Chromosomal localization of a major family of rye repeated sequences. A ^{3}H-labeled cRNA copy of the plasmid pBR322 containing part of the 480-bp highly repeated sequence from rye was hybridized to denatured metaphase chromosomes. The chromosomal sites to which the rye sequence hybridized are indicated by the silver grains. Further details on the construction of the pBR322-rye DNA plasmid are in Bedbrook et al., 1980.

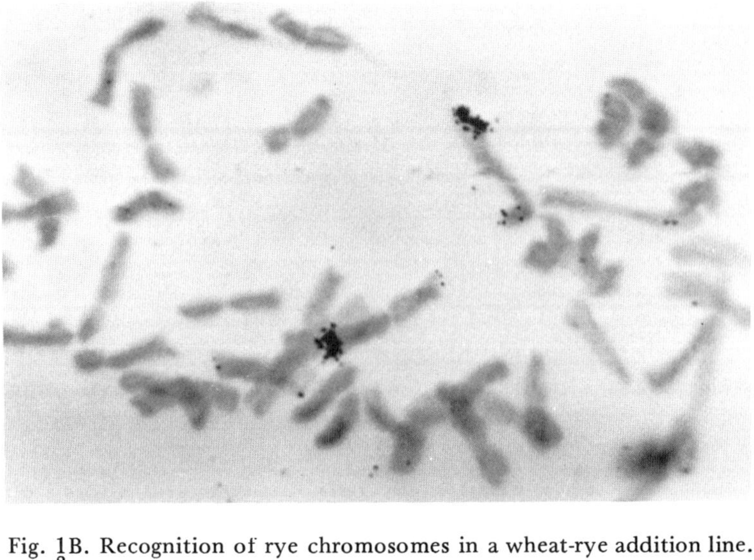

Fig. 1B. Recognition of rye chromosomes in a wheat-rye addition line. The ^{3}H-labeled cRNA used in A was hybridized to metaphase chromosomes of a plant having 21 pairs of wheat chromosomes and a single pair of rye chromosomes. Because the rye DNA sequence in the plasmid is absent from wheat, only the pair of rye chromosomes is marked by silver grains.

of probes will be isolated and characterized to give the cytogeneticist the maximum opportunity to characterize karyotypes sensitively and efficiently.

Rye Repeated Sequences and Agronomic Deficiencies in Triticale

We have already outlined some of the properties of the highly repeated sequences of rye and their value for distinguishing rye chromosomes from wheat chromosomes. The decision to purify and analyze in detail the highly repeated sequences in rye telomeric heterochromatin was also taken because of an important plant breeding problem, namely, meiotic instability and grain shriveling in triticale. We were aware through the work of colleagues in the Plant Breeding Institute that the rye heterochromatin is one component that may contribute to these aberrations.

Triticale is the name given to the hybrids formed from crossing wheat and rye (Gustafson, 1976). The possibility of combining certain attributes of rye with those of wheat in this hybrid has led plant breeders to explore intensively the creation of a new useful crop (Zillinsky, 1974). However, triticales in general suffer from sufficient meiotic irregularity, sterility, and kernel shriveling to affect seriously their economic potential.

At meiosis in triticale, the frequency of univalents is often high, and, in the cases analyzed, it is the rye chromosomes that are unpaired (Thomas and Kaltsikes, 1974, 1976). Furthermore, rye chromosome arms possessing large terminal blocks of heterochromatin are much more likely to fail to establish chiasmata in triticale than chromosome arms without heterochromatin (Thomas and Kaltsikes, 1976; Merker, 1976). Thus, it has been proposed that rye heterochromatin interferes with rye chromosome pairing or chiasma formation in triticale meiosis and is a major cause of the resulting meiotic irregularities.

Some clues to the cause of the kernel shriveling have come from cytological studies of very early endosperm development. In wheat, the first mitotic division in the egg sac contributing to the endosperm occurs 6 to 8 hours after pollination (Bennett et al., 1973).

A series of very rapid coenocytic nuclear divisions then occurs until approximately 72 hours after fertilization when cellularization occurs (Bennett, Smith, and Barclay, 1975; Kaltsikes and Roupakias, 1975). During the phase of rapid coenocytic nuclear divisions in triticale (but not in wheat and rarely in rye), abnormal nuclei are frequently visible (Bennett et al., 1973; Bennett, 1977; Kaltsikes, Roupakias, and Thomas, 1975; Kaltsikes and Roupakias, 1975). The number of these aberrant nuclei increases during the coenocytic phase of development. The most common form of aberration is polyploidy with consequential reduction in the number of separate nuclei. The importance of these observations lies in the high positive correlation between the degree of kernel shriveling at maturity and the frequency of these abnormal nuclei (Kaltsikes, Roupakias, and Thomas, 1975; Kaltsikes and Roupakias, 1975; Bennett, 1977). It is presumed that the reduced number of normal endosperm cells in regions of the kernel gives rise to localized pockets of tissue with reduced carbohydrate and protein sink capacities and that incompletely filled regions of a kernel collapse during kernel maturation.

Bennett (1977) has proposed that a substantial fraction of the aberrant polyploid nuclei have their origins in anaphase bridges that are not resolved in the coenocytic endosperm. The chromosomes in the bridges are most frequently those from rye, not wheat. One major well-known cause of anaphase bridges is failure to complete chromosome replication before anaphase. Because the terminal segments of rye chromosomes are known to be late replicating (Ayonoadu and Rees, 1973; Lima-de-Faria and Jaworska, 1972), the hypothesis has been put forward that it is the large amount of terminal heterochromatin in rye chromosomes absent from wheat chromosomes that is a major factor in endosperm shriveling in triticale. This hypothesis is supported by the observations that the frequency of aberrant nuclei in the endosperm is considerably lower in rye addition lines carrying pairs of rye chromosomes that have lost large blocks of terminal heterochromatin (Bennett, 1977). However, Darvey (1973) has shown that the enhanced frequency of aberrant endosperm nuclei is, in some cases, associated with the long arms of rye chromosomes that do not possess large blocks of heterochromatin. It is clearly important to investigate whether these chromosome arms contain late-replicating DNA segments.

Other aberrant endosperm polyploid nuclei may arise from mitotic spindle failure (Kaltsikes and Roupakias, 1975). Whether late-replicating rye DNA or heterochromatin promotes spindle failure in triticale endosperm is unknown but the rye chromosomes with telomeric heterochromatic deletions are associated with fewer polyploid nuclear aberrations.

If a major cause of kernel shriveling in triticale is late-replicating rye heterochromatin, then the unraveling of the molecular basis of the heterochromatin assumes relevance to plant breeding research. Our molecular clones of rye-specific repeated sequences, which are tandemly clustered in the telomeric heterochromatin, are now enabling us to study the structure of the heterochromatin. The plasmid-cloned probes are also enabling us to analyze specifically the sequences in the anaphase bridges. They should also allow us to help select rye lines from laboratory and/or plant breeding populations that have lost substantial amounts of terminal heterochromatin and that might be more suitable as parents for triticale. J. P. Gustafson (University of Manitoba, Winnipeg) and M. D. Bennett (Plant Breeding Institute, Cambridge) are currently constructing such triticale lines.

The consequences of rye heterochromatin in endosperm development in triticale raise the possibility that families of highly repeated sequences are not functionally inert, even though they are probably not transcribed. Although they are presumably not detrimental to rye, when they are present in the wheat genetic background, they upset the rapid endosperm mitotic and meiotic chromosome cycles. If the effect of the families of repeated sequences in rye heterochromatin on triticale meiosis and kernel development are conclusively proven, then they may constitute the first nucleotide sequences to be characterized and cloned in bacterial plasmids that are not transcribed but have substanial economic consequences on crop performance in agriculture.

Conclusions

The studies we have outlined in this paper are examples of the application of DNA biochemistry and molecular biology to cereal chromosome cytogenetics, which has been part of plant breeding research for decades. The DNA studies are not of a nature to provide revolutionary quantum leaps in plant breeding in the foreseeable future. However, it may well be that the most likely way that molecular biology will contribute to plant breeding advances in the next few years is by bringing greater understanding, efficiency, and precision to the processes and practices that plant breeders currently use. If molecular biology is to be useful to plant breeding, it surely must be developed alongside plant breeding research. Without the close association, the fruits of molecular biology for economic improvements in crop performance will inevitably take a much longer time to ripen.

References

Appels, R.; Driscoll, C.; and Peacock, W. J. (1978) Heterochromatin and highly repeated DNA sequences in rye (*Secale cereale*). *Chromsoma (Berlin) 70*, 67-89.

Ayonoadu, U., and Rees, H. (1973) DNA synthesis in rye chromosomes. *Heredity 30*, 233-40.

Bedbrook, J.; Gerlach, W.; Thompson, R.; Jones, J.; and Flavell, R. (1980) Molecular cloning of higher plant DNA. In this volume.

Bennett, M. D. (1977) Heterochromatin, aberrant endosperm nuclei and grain shrivelling in wheat-rye genotypes. *Heredity 39*, 411-19.

Bennett, M. D.; Rao, M. K.; Smith, J. B.; and Bayliss, M. W. (1973) Cell development in the anther, the ovule and the young seed of *Triticum aestivum* L. var. Chinese Spring. *Phil. Trans. Roy. Soc. Lond. B. 266*, 39-81.

Bennett, M. D.; Smith, J. B.; and Barclay, I. (1975) Early seed development in the Triticeae. *Phil. Trans. Roy. Soc. Lond. B. 272*, 199-227.

Bennett, M. D.; Gustafson, J. P.; and Smith, J. B. (1977) Variation in nuclear DNA in the genus *Secale*. *Chromosoma (Berlin) 61*, 149-76.

Bhatia, C. R., and Smith, H. H. (1966) Variation associated with an *Aegilops umbellulata* chromosome segment incorporated into wheat. *Nature 211*, 1425-26.

Chapman, V., and Miller, T. E. (1978) The amphiploid of *Hordeum chilense* X *Triticum aestivum*. *Cereal Res. Commun. 6*, 351-52.

Darvey, N. L. (1973) Genetics of seed shrivelling in wheat and triticale. In *Proc. 4th Internat. Wheat Symp.*, pp. 155-59. University of Missouri, Columbia.

Darvey, N. L., and Gustafson, J. P. (1975) Identification of rye chromosomes in wheat-rye addition lines and triticale by heterochromatin bands. *Crop Sci. 15*, 239-43.

Driscoll, C. J., and Jensen, N. F. (1963) A genetic method for detecting intergeneric translocation. *Genetics 48*, 459-68.

Driscoll, C., and Sears, E. R. (1971) Individual addition of the chromosomes of "Imperial" rye to wheat. *Agron. Abst.*, 6.

Dvorak, J., and Knott, D. R. (1974) Disomic and ditelosomic additions of diploid *Agropyron elongatum* chromosomes to *Triticum aestivum. Can. J. Genet. Cytol. 16*, 399-417.

Flavell, R. B.; Rimpau, J.; and Smith, D. B. (1977) Repeated sequence DNA relationships in four cereal genomes. *Chromosoma (Berlin) 63*, 205-22.

Flavell, R. B.; O'Dell, M.; Rimpau, J.; and Smith, D. B. (1978) Biochemical detection of alien DNA incorporated into wheat by chromosome engineering. *Heredity 40*, 439-55.

Gerlach, W. L.; Appels, R.; Dennis, E. S.; and Peacock, W. J. (1978) Evolution and analysis of wheat genomes using highly repeated DNA sequences. *Proc. 5th Internat. Wheat Genet. Symp.*, pp. 81-91. New Delhi, India.

Gustafson, J. P. (1976) The evolutionary development of triticale: The wheat-rye hybrid. *Evol. Biol. 9*, 107-35.

Hart, G. E.; McMillin, D. E.; and Sears, E. R. (1976) Determination of the chromosomal location of a glutamate oxaloacetate transaminase structural gene using *Triticum agropyron* translocation. *Genetics 83* 49-61.

Islam, A. K. M. R.; Shepherd, K. W.; and Sparrow, D. H. B. (1975) Addition of individual barley chromosomes to wheat. *Proc. 3rd Internat. Barley Genet. Symp.*, pp. 260-70.

Kaltsikes, P. J., and Roupakias, D. B. (1975) Endosperm abnormalities in *Triticum-Secale* combinations. II. Addition and substitution lines. *Can. J. Bot. 53*, 2068-76.

Kaltsikes, P. J.; Roupakias, D. G.; and Thomas, J. B. (1975) Endosperm abnormalities in *Triticum-Secale* combinations. I: Triticosecale and its parental species. *Can. J. Bot. 53*, 1050-67.

Knott, D. R. (1961) The inheritance of rust resistance. VI: The transfer of stem rust resistance from *Agropyron elongatum* to common wheat. *Can. J. Plant Sci. 41*, 109-23.

Lima-de-Faria, A., and Jaworska, H. (1972) The relationship between chromosome size gradient and the sequence of DNA replication in rye. *Heredity 70*, 39-58.

Merker, A. (1976) The cytogenetic effect of heterochromatin in hexaploid triticale. *Hereditas 83*, 215-22.

Riley, R., and Chapman, V. (1958) The production and phenotypes of wheat-rye chromosome addition lines. *Heredity 12*, 301-5.

Riley, R., and Chapman, V. (1960) The meiotic behaviour, fertility and stability of wheat-rye chromosome addition lines. *Heredity 14*, 89-100.

Riley, R.; Chapman, V.; and Johnson, R. (1968) The incorporation of alien disease resistance in wheat by genetic interference with the regulation of meiotic chromosome synapsis. *Genet. Res. (Cambridge) 12*, 199-219.

Riley, R., and Kimber, G. (1966) The transfer of alien genetic variation to wheat. *Ann. Rep. Plant Breeding Institute, Cambridge.*

Rimpau, J.; Smith, D. B.; and Flavell, R. B. (1978) Sequence organisation analysis of the wheat and rye genomes by interspecies DNA/DNA hybridisation. *J. Mol. Biol. 123*, 327-59.

Sears, E. R. (1956) The transfer of leaf-rust resistance from *Aegilops umbellulata* to wheat. *Brookhaven Symp. Biol. 9*, 1-22.

Sears, E. R. (1972) Chromosome engineering in wheat. *Stadler Symp. 4*, 23-38.

Singh, R. J., and Robbelen, G. (1976) Giemsa banding technique reveals deletions within rye chromosomes in addition lines. *Z. Pflanzenzüchtg. 76*, 11-18.

Thomas, J. B., and Kaltsikes, P. J. (1974) A possible effect of heterochromatin on chromosome pairing. *Proc. Nat. Acad. Sci. U.S.A. 71*, 2787-90.

Thomas, J. B., and Kaltsikes, P. J. (1976) The genomic origin of the unpaired chromosomes in triticale. *Can. J. Genet. Cytol. 18*, 687-700.

Zillinsky, F. J. (1974) The development of triticale. *Adv. Agron. 26*, 315-48.

Part III: Gene Isolation and Transfer

Molecular Cloning of Higher Plant DNA

J. Bedbrook, W. Gerlach, R. Thompson, J. Jones, and R. B. Flavell

In vitro recombinant DNA technology makes possible the isolation and large-scale preparation of specific DNA fragments from the chromosomes of higher organisms. It therefore makes possible molecular analysis of eukaryotic genes and chromosomes. The success of this technology (referred to here as molecular cloning) has already shown itself in work with fungal, insect, and higher animal DNAs. Though molecular cloning has proven productive in studying chloroplast genes (Bedbrook et al., 1979d; Coen et al., 1978), many workers have encountered problems when attempting to clone nuclear DNA sequences from various plants. In this paper, we will attempt to rationalize some of the problems associated with the cloning of plant DNA. From the results of our experiments with cereal plants, it is possible to evaluate some of the problems associated with the molecular cloning of plant DNA.

Note: J. Bedbrook is a long-term European Molecular Biology Organization fellow. W. Gerlach was the recipient of a Commonwealth Scientific and Industrial Research Organization postdoctoral studentship. All recombinant DNA experiments described in this paper were carried out under class II containment as defined by the Genetic Manipulation Advisory Group of the United Kingdom.

Some General Considerations

DNA Purity

Preparation of high-molecular-weight, pure DNA is a prerequisite for efficient cloning experiments. For cereals, we first isolate nuclei from embryo tissue on metrizamide gradients. Nuclei are then lysed with ionic detergents and the DNA purified by two or three cycles of centrifugation in CsCl-ethidium bromide gradients. The inclusion of saturating amounts of ethidium bromide in these gradients enables the DNA to be separated from major polysaccharide containments.

Nuclear DNA Content and Repeated DNA Sequences

The probability of cloning any single DNA sequence is inversely related to genome size. The DNA contents of several common experimental plants are given in Table 1. Those of *Drosophila melanogaster* and mouse are given for comparison. Most higher plants have relatively large DNA contents.

In the high-C-value plants, most of the DNA is in the form of repeated DNA sequences existing in blocks of tandem arrays or

Table 1. DNA Content and 5-Methyl Cytosine Content in Various Species.

Species	pg DNA/2C cell	$\% \frac{5MC^*}{C+5MC}$
Barley	10.9	23.1
Carrot	2.0	25.7
Cotton		23.2
Pea	10.5	23.8
Rye	16.6	26.7
Spinach		20.4
Tobacco	7.8	24.6
Wheat	31.4	25.0
Maize	4.7	26.7
D. melanogaster	0.17	< 0.1
Mouse	6.0	5.0

*Percentages of total cytosine in the form of 5-methyl cytosine.

dispersed throughout the chromosome (Flavell et al., 1974). A generalized view of the size and arrangement of repeated sequences in plants has been provided by various kinds of solution hybridization studies (e.g., Rimpau et al., 1978). Very little specific information on the nucleotide sequence of this repeated DNA is available. There is evidence (Brutlag et al., 1977) that certain simple repeated sequences in *D. melanogaster* are unstable when cloned in *Escherichia coli*. This raises the possibility that much of the large fraction of repeated DNA from plants may be unstable in *E. coli*.

Methylated Cytosine

Most plants contain a considerable proportion of 5-methyl cytosine (5MC) in their chromosomal DNA. The proportion of total cytosine in the form of 5MC is given for a range of species in Table 1. Methylated cytosine may be handled differently from cytosine by *E. coli* DNA-restriction and DNA-recombination systems. 5MC may, therefore, reduce the probability of certain plant DNA sequences being faithfully propagated in *E. coli*. Another important point is that 5MC in specific sequence locations prevents certain restriction endonuclease from cutting DNA. Such enzymes may, therefore, be unsuitable for use in the molecular cloning of plant DNA since the size of many of the resulting DNA fragments is prohibitively large for successful transformation of *E. coli* cells.

Restriction Enzyme Digestion of Plant DNA

Molecular cloning of DNA fragments produced by restriction endonucleases is convenient since it enables rapid identification of the cloned sequence. Fig. 1 shows the fragments produced by digestion of rye nuclear DNA with various enzymes fractionated by gel electrophoresis. Bands superimposed on the smear of fluorescence arise from digestion of families of repeated DNA sequences. Hamer and Thomas (1974) have shown that the size distribution of DNA fragments produced by restriction-enzyme digestion of

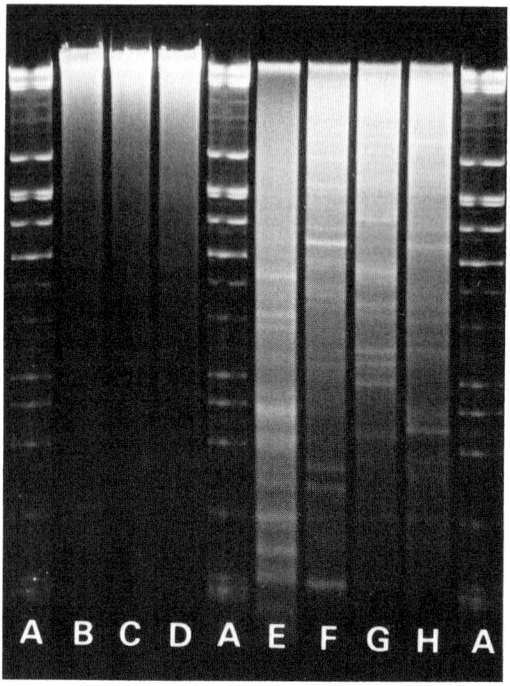

Fig. 1. Agarose gel electrophoresis of rye DNA digested with various restriction endonucleases. Rye DNA (5 μg) prepared from nuclei by two cycles of density gradient centrifugation in CsCl-ethidium bromide was digested to completion with restriction endonucleases. Digests were fractionated by electrophoresis through 1% agarose gels. (A) Size markers, (B) rye DNA digested with Sma II, (C) Hpa II, (D) Pst I, (3) Hind III, (F) Bgl II, (G) BamH I, (H) Eco RI. Electrophoresis was from top to bottom.

Drosophila melanogaster DNA fits that predicted by random fragmentation theory. Analysis of the relationship between fragment size and the integral of the amount of DNA over the range of sizes produced by restriction enzymes enables calculation of the number average size of DNA fragments produced by the enzyme. This is a useful value since it enables the calculation of the total number of particular restriction fragments in the genome.

Correspondence of the theoretical and determined average sizes is indicative of a predominantly random distribution of nucleotides throughout the chromosomes.

Table 2 gives the theoretical and observed number average sizes of rye DNA fragments produced by various restriction enzymes. The observed number average fragment size values after restriction with Sma I and Hpa II are considerably higher than predicted. This is almost certainly due to the extensive C methylation in rye DNA since these enzymes do not cut DNA if the C residue 5′ to the cutting site is methylated. Interestingly, the enzyme Hae III, which does not cut if the C residue 3′ to the cutting site (Mann and Smith, 1977) is methylated, gives a number average molecular weight close to the theoretical value. Taken together, these data suggest that most methylated cytosines in plant DNA exist in the sequence C*p G as is true for animal DNA (Bird and Southern, 1978; Bird, 1978). The observed number average molecular weight for rye DNA fragments produced by the enzyme Pst I is significantly higher than the theoretical value for an enzyme with the Pst recognition sequence. We can explain this result only by suggesting that this sequence is underrepresented in the rye genome. Though the other enzymes listed in Table 2 give number average molecular weight values close to theoretical values, the distribution of the size classes produced by all the enzymes listed, with the exception of Hae III, is not that expected from a random arrangement of restriction sites. The numbers of fragments considerably larger and considerably smaller than the number average fragment molecular weight are greater than expected for a random nucleotide sequence.

Table 2 also gives an estimate of the total number of restriction fragments produced by the various enzymes. With enzymes suitable for cloning purposes—such as Eco RI, Bam HI, Hind III, Bgl II—the number of fragments produced is more than 10^6. These very large numbers make total genome cloning, after complete digestion with restriction enzymes, prohibitive with high-C-value plants.

Maniatis and his coworkers (1978) have presented schemes for preparing clone libraries of DNA fragments from partial digests. For rye DNA, partial digests giving average insert lengths of 15Kb would still require some 6×10^5 independent clones to give a 50% chance of seeing every unique sequence.

Table 2. Comparison of Theoretical and Empirical Frequencies for the Occurrence of Restriction Endonuclease Targets in the DNA of *Secale cereale*

Restriction Enzyme	Recognition Sequence	Probability of Occurrence of Sequence in Genome p	Mean Segment Length Determined from p	Calculated Mean Segment Length Determined from Restriction Enzyme Digests	Calculated Number of "Different" Fragments Produced by Digestion of *S. cereale* DNA
EcoR I	G′ AA*TTC	2.887×10^{-4}	3464	3400	2.30×10^{6}
Bam HI	G′GATC*C	1.938×10^{-4}	5132	3311	2.35×10^{6}
Bgl III	A′GATCT	2.887×10^{-4}	3464	2512	3.10×10^{6}
Hae III	GG′C*C	25.860×10^{-4}	387	400	1.95×10^{7}
Hpa II	CC*′GG	25.860×10^{-4}	387	>10,000	$< 7.80 \times 10^{5}$
Hind III	A*′AGCTT	2.887×10^{-4}	3464	2,400	3.25×10^{6}
Pst I	CTGCA′G	1.948×10^{-4}	5132	8,128	9.50×10^{5}
Sma I	CCC*′GGG	1.315×10^{-4}	7605	>10,000	$< 7.80 \times 10^{5}$

Note: Rye DNA was digested with restriction enzymes and fractionated by gel electrophoresis as described in the legend to Fig. 1. Traces of photographic negatives of the gel were made using a scanning densitometer. Cumulative size distributions were obtained by measuring the fluorescence within specific fragment-size classes determined from molecular weight markers. The integral of the weight over the range of fragment sizes was plotted against fragment size. The plots were tested for conformance to the Kuhn approximation ($w_t \cong tp^2 e^{-pt}$) where w_t is the weight fraction of DNA found in segments t base pairs long. The value for p (probability of a double-strand cleavage) was determined from the base composition of rye DNA and the base composition of the enzyme recognition sequence.

General Observations on the Cloning Efficiency of Cereal DNA

Table 3 shows results of experiments in which the relative cloning efficiency of wheat DNA and *E. coli* DNA prepared either by CsCl-ethidium bromide or by CsCl-ethidium bromide plus CsCl-actinomycin-D density gradient centrifugation were compared. In this experiment, closed circular vector plasmid DNA gave 10^6 transformants per microgram of DNA. This number was reduced by 3 orders of magnitude when the plasmid DNA was linearized with EcoR I. Self-ligation of the linear plasmid increased the number of transformants to 10^5 per microgram of DNA. The number of transformants given by each DNA sample was standardized by using an internal plasmid DNA control (see note to Table 3). Ligation of *E. coli* DNA, prepared by either method, did not reduce the number of transformants given by self-ligated pAC184 plasmid DNA. However, ligation of wheat DNA prepared by CsCl-ethidium bromide gradients to pAC184 did reduce the number of transformants produced by 1 μg of this plasmid by a factor of 4, and ligation of wheat DNA prepared by CsCl-ethidium bromide plus CsCl-actinomycin-D reduced the transformant number by a factor of 2.5. The frequency of *E. coli* and wheat DNA inserts in pAC184 was about the same. The reduction in transformation frequency by ligation to wheat DNA is a consistent feature for which we have no simple explanation. We have not yet made a systematic study of the size of inserts in clones of wheat DNA, but we have obtained clones with inserts ranging in length from 15,000 bp to less than 200 bp with Eco RI digests with no bias for large or small inserts relative to the number average length of Eco RI fragments. None of the characterized chimeric plasmids have shown instability on storage, although a small fraction of cloned sequences appear to contain deletions of the natural sequence.

Cloning of Specific Sequences from Cereals

In this section, we discuss results of cloning experiments designed to isolate and characterize specific DNA fragments from cereals.

Table 3. The Relative Cloning Efficiency of *E. coli* and wheat DNA

DNA	Number of Transformants per μg Vector DNA	Frequency Transformants per Cell	% Inserts
None	0	0	–
Closed circular pAC184 DNA	1.0×10^6	1.0×10^{-3}	–
EcoR I digested pAC184 DNA	1.2×10^3	1.2×10^{-6}	–
Ligated pAC184 DNA	1.0×10^5	1.0×10^{-4}	–
pAC184 DNA ligated to wheat DNA 1	2.5×10^4	2.5×10^{-5}	11
pAC184 DNA ligated to wheat DNA 2	4.0×10^4	4.0×10^{-5}	12
pAC184 DNA ligated to E. coli DNA 1	8.5×10^4	8.5×10^{-5}	13
pAC184 DNA ligated to *E. coli* DNA 2	1.2×10^5	1.2×10^{-4}	12

Note: Wheat DNA sample 1 (10 μg) was purified by two cycles of CsCl-EtBr density gradient centrifugation. Wheat DNA sample 2 (10 μg) was purified by one cycle of CsCl-EtBr density gradient centrifugation followed by one cycle of CsCl-Act-D density gradient centrifugation. *E. coli* DNA sample 1 (10 μg) was purified by two CsCl-EtBr gradients and *E. coli* DNA sample 2 (10 μg) was purified by CsCl-EtBr gradient centrifugation followed by CsCl-Act-D gradient centrifugation. The DNAs were digested to completion with Eco RI and ligated individually with 0.2 μg of Eco RI digested pAC184 plasmid DNA. The ligated DNA was mixed with 0.1 μg of untreated RSf1030 plasmid DNA. RSf1030 and pAC184 are compatible plasmids (Bedbrook, Lehrach, and Ausubel; manuscript submitted). The DNA samples were used to transform 10^9 $CaCl_2$-treated *E. coli* HB101 cells. As controls closed circular pAC184 DNA (0.2 μg), Eco RI-digested pAC184 DNA (0.2 μg) and pAC184 DNA digested with EcoR I and self-ligated were also used to transform HB101. The cells were plated independently on media containing 30 μg/ml ampicillin, to select RSf1030 transformants, on media containing 10 μg/ml tetracycline to select pAC184 transformants, and on unselective media to assay the total number of cells per transformation. The number of pAC184 transformants was standardized for each transformation using the number of RSf1030 transformants. The frequency of transformation was standardized using the number of colonies on unselective plates. The fraction of pAC184 transformants containing DNA fragments inserted at the EcoR I site was determined by testing for sensitivity to chloramphenicol.

Cloning Highly Repeated DNA Fragments from Rye

IDENTIFICATION OF REPEATED SEQUENCES OF RYE TELOMERIC HETEROCHROMATIN. Rye chromosomes are characterized by the presence of blocks of heterochromatin at the telomeres; various taxa of rye have different amounts of such heterochromatin (Bennett et al., 1977). We sought to determine the nature of the DNA sequences in these telomeres. In known cases, for example, the centromeres of *Drosophila melanogaster* (Peacock et al., 1978), the DNA of heterochromatin has been found to consist of a few highly repeated sequence DNA families. In many species, these repeated DNA sequences can be isolated as density satellites. In rye, we have not been able to isolate density satellites accounting for much of the repeated sequence fraction. To approach characterization of the telomere sequences in rye, we compared restriction-enzyme digests fractionated on gels of two species that have vastly different amounts of telomeric heterochromatin. These comparisons were made in an attempt to find repeated DNA bands associated with the telomeres. Fig. 2 shows a scan of the fluorescence resulting from a Hind III digest of cultivated rye (*S. cereale*),

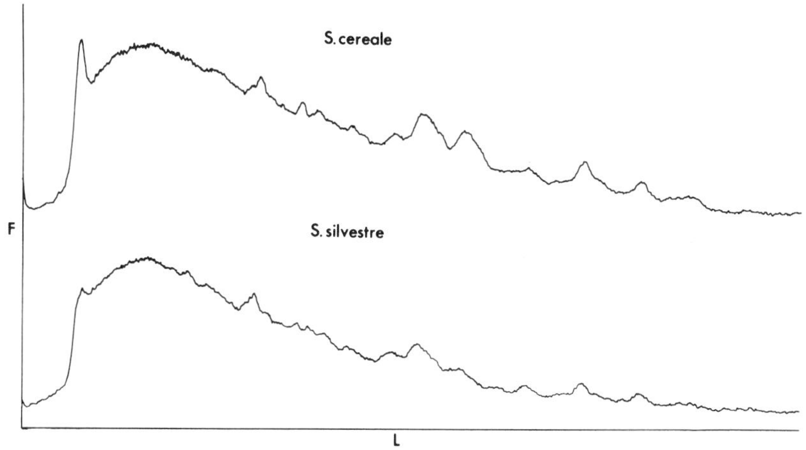

Fig. 2. Densitometer tracings of rye and *S. silvestre* DNA digested with Hind III and fractionated on a 1% agarose gel. Rye and *S. silvestre* DNAs were prepared, digested, and fractionated as described in the caption of Fig. 1. Photographic negatives of the stained gels were scanned in a densitometer. Electrophoresis was from left to right.

which has large telomeric heterochromatic blocks, and *S. silvestre*, which has small telomeric heterochromatic blocks (Bennett et al., 1977). The major difference between the digests is that *S. cereale* DNA contains considerably more fluorescence at the limiting mobility in the gel electrophoresis. The DNA at the limiting mobility (which we call the Hind III relic DNA, see Fig. 1E) has a number average length approximately two times the length of phage λ DNA. Relatively pure fractions of this relic DNA could be obtained from Hind III digests of rye DNA by velocity sedimentation in sucrose density gradients. Fig. 3 shows Hae III digests of *S. cereale*.

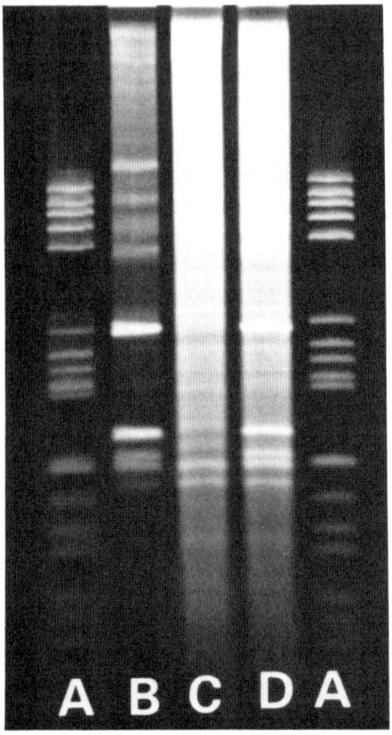

Fig. 3. Gel electrophoresis of Hae III digests of rye and *S. silvestre* DNA. Rye and *S. silvestre* DNA were prepared as described in the caption of Fig. 1. Rye DNA fragments uncut by Eco RI (Eco RI relic) were fractionated from Eco RI-restricted DNA by velocity sedimentation on sucrose gradients. Digests of the DNA were fractionated by electrophoresis on a 6% polyacrylamide gel. (A) pBR322 DNA cut with Hae III, (B) Eco RI relic (rye) digested with Hae III, (C) *S. silvestre* DNA digested with Hae III, (D) rye DNA digested with Hae III.

DNA (3D) and *S. silvestre* DNA (3C). Several major bands are present in digests of both species; however, two major visible bands are present only in *S. cereale* DNA digests. Hae III digestion of sucrose gradient purified Hind III relic DNA is shown in Fig. 3B. Hind III relic DNA contains, as predicted from the data shown in Fig. 2, DNA sequences that are unique to *S. cereale* DNA. This DNA fraction also contains major bands common to the two species.

Preliminary *in situ* hybridization with probes made from Hind III relic preparations suggested that most of the sequences present in this fraction were located at the telomeres of rye chromosomes.

MOLECULAR CLONING OF THE REPEATED SEQUENCES OF RYE TELOMERES: To obtain a detailed molecular description of the telomeres of rye, we selected clones of sequences identified to reside at the telomeres. Fig. 4 illustrates the cloning scheme. Total rye chromosomal DNA was digested with Hae III, ligated to the "linker" sequence $^{5'}$CCAAGCTTCC$^{3'}$, which contains the recognition sequence for

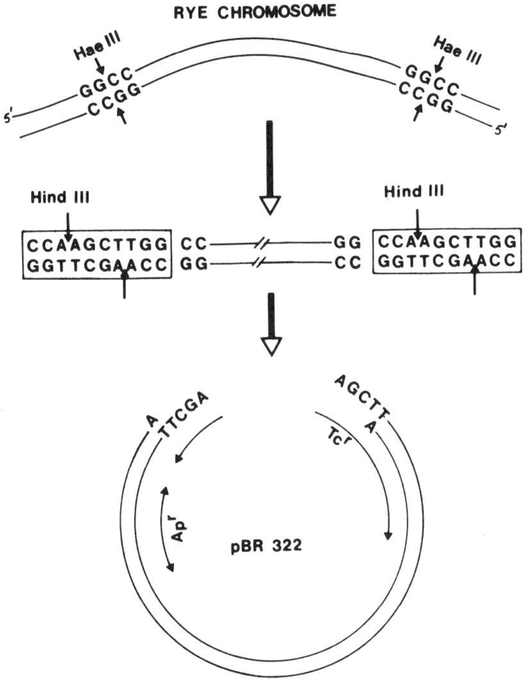

Fig. 4. Scheme for the cloning of the Hae III fragments of rye DNA. Rye DNA (20 μg), digested to completion with Hae III, was ligated to 2 μg of a decanucleotide "linker" sequence containing the recognition sequence for Hind III. The ligated DNA was digested with Hind III and ligated to Hind III digested pBR322 DNA (1 μg). The DNA from the second ligation was used to transform the *E. coli* strain HB101 to ampicillin resistance. Transformants containing chimeric plasmids were screened by testing for tetracycline sensitivity.

Hind III. After digestion with Hind III, this DNA was ligated to Hind III digested pBR322 plasmid DNA. Transformants containing inserted rye DNA were selected in the *E. coli* strain HB101.

Hind III relic DNA, prepared as described earlier, was digested with Hae III, fractionated on acrylamide gels and all the bands visible in Fig. 3B were excised and the DNAs purified. ^{32}P labeled cRNAs were prepared using the various Hind III relic DNA Hae III fragments as templates. These cRNAs were individually hybridized to 200 clones from the cloning experiment described earlier, using the method of Grunstein and Hogness (1975). A total of only five families of Hae III clones was obtained from hybridization of all the probes. Thus, several of the probes prepared from different bands hybridized the same clones, suggesting that different bands could be different configurations of the same sequence. Fig. 5 shows Hae III digests of several members of one of the five clone families obtained. Fig. 5G shows a Hae III digest of pBR322 DNA an 5H shows a Hae III digest of total rye DNA. Fig. 5A through 5F show six clones hybridizing the major band in the Hae III digest of rye (5H). Though all the various clones contain several bands not found in vector digests, all the clones contain a band comigrating with the major Hae III band from rye DNA used as the probe to select this particular family.

ORGANIZATION OF THE DNA SEQUENCES IN THE TELOMERES OF RYE CHROMOSOMES. A detailed description of the structure and location of telomere repeated sequences in rye is reported elsewhere (Bedbrook et al., 1979b; Flavell et al., 1980). Briefly, the use of the clone families, isolated as described earlier, has enabled us to determine the nature of the bulk of the DNA sequences in the telomeres of rye chromosomes. The five clone families are derived from four families of highly repeated sequences arranged in tandem arrays. Fig. 6 gives line drawings of the repeats. Family 1, which is present only in *S. cereale* and absent from *S. silvestre,* consists of tandem repeats, defined by Mbo II digestion with a repeating unit length of 480 nucleotides. This repeat accounts for two of the five Hae III clone families. Family 1 comprises approximately 5% of the total DNA of *S. cereale* and is present on the ends of all seven chromosomes.

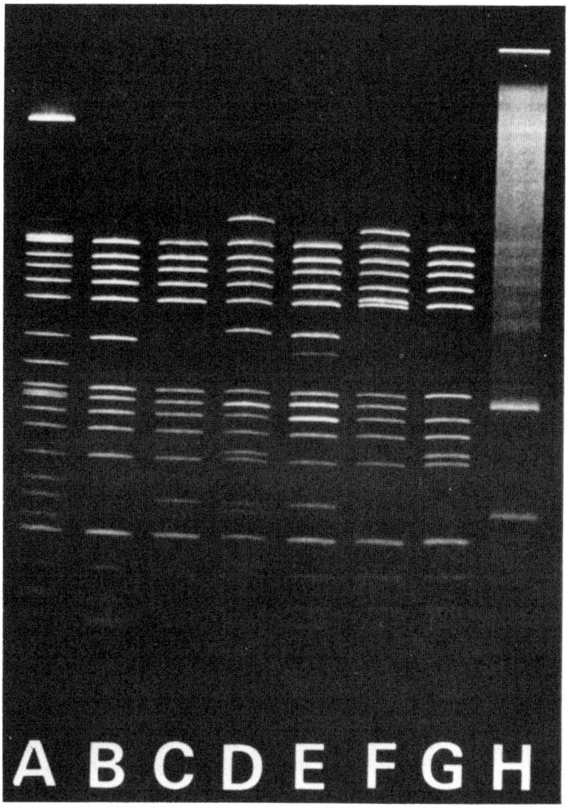

Fig. 5. Gel electrophoresis of molecular clones of Hae III fragments of rye DNA. Plasmid DNA, containing rye Hae III fragment inserts, and total rye DNA were digested with Hae III and fractionated by gel electrophoresis on a 6% polyacrylamide gel. (A-F) Hae III digests of chimeric plasmid DNAs containing sequences that hybridized to the major repeated DNA fragment in rye. (G) pBR322 DNA digested with Hae III. (H) rye DNA digested with Hae III.

Family 2, like family 1, is present only in *S. cereale* and is absent from *S. silvestre*. Mbo II digests show a basic repeat unit of 600 nucleotides. However, Hae III digests suggest a subrepeat of approximately 110 nucleotides. This family comprises approximately 2.5% of *S. cereale* DNA and is present exclusively at the telomeres of six of the seven *S. cereale* chromosomes.

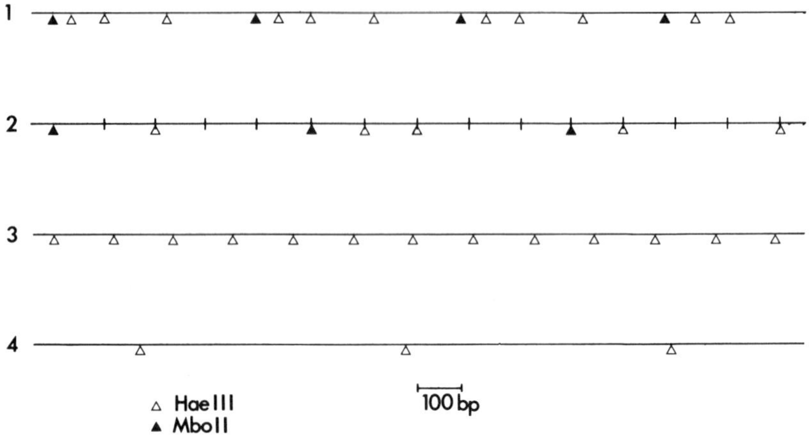

Fig. 6. Restriction endonucleases maps for the major DNA repeats in telomeric heterochromatin of rye. Lines 1-4 show tandem arrays of the major repeats in rye telomeric heterochromatin. Triangles indicate sites for restriction endonucleases.

Family 3 is defined by Hae III digestion to have a basic repeat unit of 120 nucleotides arranged in tandem arrays. This family is present in both *S. cereale* and *S. silvestre* and represents approximately 2% of *S. cereale* DNA. Though most of the DNA of this family is located at the telomeres of *S. cereale* and *S. silvestre*, there are also major interstitial locations on many chromosomes for sequences in this family.

Family 4 consists of tandem arrays of repeats approximately 620 nucleotides. There is evidence that this sequence, as for the repeat of family 2, has a simpler subrepeating length. Family 4 represents less than 0.5% of total *S. cereale* DNA.

The way in which the different repeats are organized in relation to one another is unknown.

Cloning of Ribosomal RNA Genes from Wheat and Barley DNA.

In experiments not illustrated here, we found that the ribosomal DNA (rDNA) repeat lengths of wheat and rye DNA were close to

9 Kb and contained a single site for the restriction endonuclease EcoR I. This value is close to that determined by Appels and his associates (1979). These experiments also indicated some length heterogeneity in the rDNA. In order to determine the basis of this length heterogeneity as well as to obtain a more detailed picture of rDNA in wheat and barley, we have cloned rDNA of these species.

ENRICHMENT FOR rDNA: In view of the size of cereal genomes and in view of the fact that we hoped to clone relatively minor variants of rDNA from cereals, we considered it worthwhile to enrich total DNA preparations for rDNA. To do this, we used the actinomycin D-CsCl gradient technique described by Hemleben and Grierson (1978). Fig. 7 illustrates results of such fractionations for wheat (7A and 7B) and for barley (7C). DNA-hybridizing radioactive rRNA is located at a low density relative to main band DNA. These gradients gave fractions enriched more than 30 times for rDNA. Gel electrophoresis fractionation of restriction endonuclease digests of ribosomal gene enriched DNA is shown in Fig. 8. For both barley DNA (8C) and wheat DNA (8F), the ribosomal DNA bands are the major bands in the digests.

CLONING OF rDNA: DNA enriched for rDNA as described earlier was digested with EcoR I and ligated to the plasmid pAC184 and used to transform *E. coli* strain HB101. Table 4 summarizes the results of the transformation. The fraction of clones containing ribosomal DNA is close to the number average fraction of the starting DNA fragments in the form of full-length rDNA repeats.

Fig. 9 shows gel electrophoresis of all rDNA plasmids obtained from the cloning experiment described in the note on Table 4. Wheat rDNA inserts (b-f) are the same size as that of the "natural" rDNA repeat (a). Twelve of the 13 barley ribosomal gene inserts contain full-length repeats (h-r). Three (i, k, and p) contain the shorter repeat while the remainder contain the longer version of the repeat (see Fig. 8C) found in Eco RI digests of natural barley rDNA. Fig. 9s shows the only clone that is not a full-length repeat insert. This plasmid contains a deletion in both the rDNA and pAC184 DNA portions of the chimeric plasmid. It is possible that this plasmid transformed the recipient strain as a linear chimera and underwent deletion prior to circularization in the host.

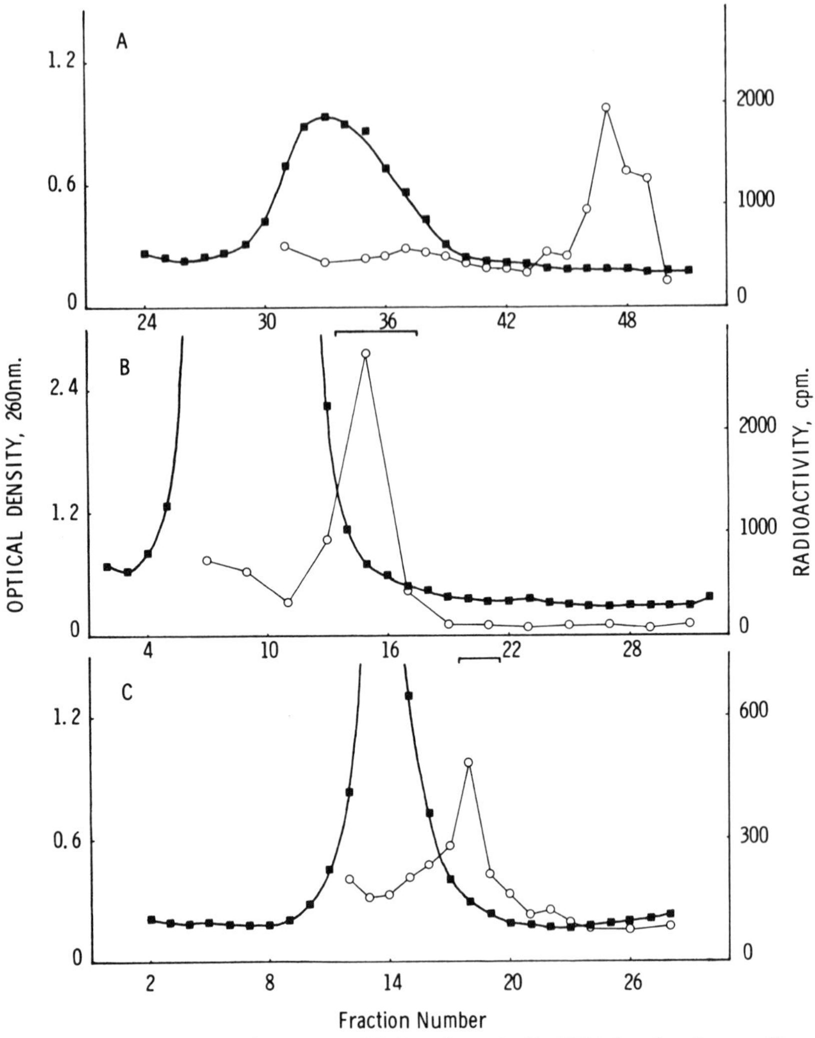

Fig. 7. Fractionation of ribosomal DNA from bulk DNA by density gradient centrifugation in CsCl-actinomycin D. DNA prepared either from nuclei or total etiolated shoot tissue was dissolved with CsCl (p = 1.70 g/cm^3) plus actinomycin D in 25 m*M* borate pH 9.2 at the rate of 0.5-1:1 w/w act-D: DNA. Samples were centrifuged at 30,000 rpm for 3 days at 3° C. The gradients were fractionated and the fractions assayed for optical density and hybridization to ribosomal RNA. F1 is the bottom of the gradient.■ Optical density at 260 nm. ○ cpm hybridized to ribosomal RNA. (A) Gradient of high molecular weight wheat DNA from nuclei. (B) Gradient of shoot DNA from barley. (C) Gradient of shoot DNA from wheat.

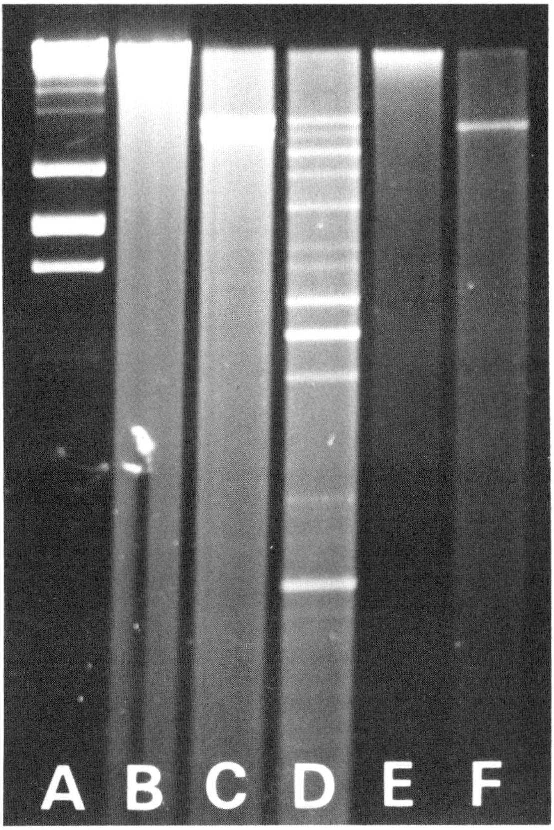

Fig. 8. Gel electrophoresis of wheat and barley DNAs enriched for ribosomal DNA. DNA fractions enriched for rDNA were prepared as described in the legend to Fig. 7. DNA (1 μg) was digested with restriction endonucleases and fractionated by electrophoresis on a 1% agarose gel. (a) λ phage DNA digested with Eco RI, (B) undigested barley rDNA, (C) barley rDNA digested with Eco RI, (D) barley rDNA digested with Bam HI, (E) undigested wheat rDNA, (F) wheat rDNA digested with Eco RI.

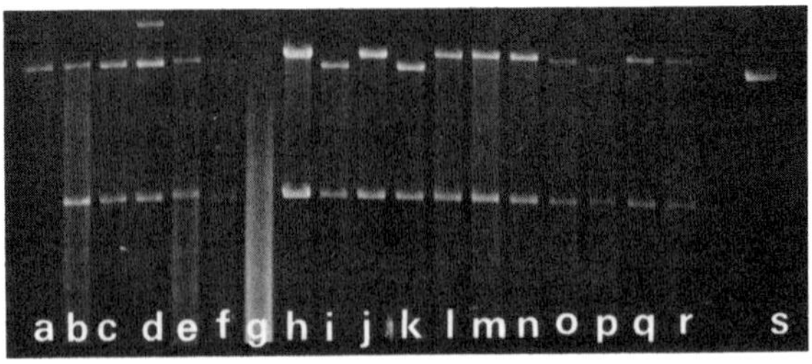

Fig. 9. Gel electrophoresis of plasmid DNAs containing sequences complementary to ribosomal RNA of wheat or barley. Plasmid DNAs from cloning Table 4 containing DNA-hybridizing ribosomal RNA were digested with EcoR I and fractionated by electrophoresis through 1% agarose gels. (a) Wheat DNA enriched for rDNA digested with Eco RI; (b-f) plasmid DNAs containing wheat rDNA digested with Eco RI; (g) barley DNA enriched for rDNA digested with Eco RI (this DNA sample was degraded prior to digestion); (h-s) plasmid DNAs containing barley rDNA digested with Eco RI.

Table 4. Cloning of rDNA from Wheat and Barley

rDNA	Fraction DNA[a] in Full-Length rDNA Repeat	Transformation[b] Efficiency	Insert Frequency	Fraction of Inserts Containing Full-Length rDNA Repeat
Wheat	1.12%	6×10^4	4.9%	1.8%
Barley	2.01%	2.4×10^4	12.1%	3.6%

Note: Wheat DNA (7.5 μg) and barley DNA (11.6 μg) enriched for rDNA as described in the legend to Fig. 7 was digested with Eco RI and ligated to 0.5 μg of pAC184 DNA. Ligated DNA was used to transform the *E. coli* strain HB101. Transformants were selected for resistance to tetracycline, and clones containing inserts were screened by sensitivity to chloramphenicol. Clones containing the sequences complementary to rRNA were screened using the procedure described by Grunstein and Hogness (1975).

[a]Number average fraction Eco RI digests in the form of full-length rDNA repeat units as determined by scans of agarose gels of the digests.
[b]Number of transformants per μg of vector DNA.

PHYSICAL MAPS OF VARIOUS rDNA REPEATS: As mentioned earlier (Fig. 8C), Eco RI digestion defines two size classes of rDNA in barley. Eco RI + Bam HI digestion of wheat DNA define three size classes of rDNA in this species (Gerlach and Bedbrook, 1979). We have cloned examples of both barley classes and two of the three wheat classes. Fig. 10 gives schematic diagrams for the four different cloned repeats. The line joining the Eco RI sites shows the extent of the cloned fragments for each example. Fig. 10A shows the major rDNA repeat of wheat and 10B is representative of one of the two identifiable minor classes. The major repeat class is 8.8 kb. The nontranscribed spacer is 3 Kb and spacer between the 18S and 25S coding sequence is 1 Kb. The length of DNA fragments

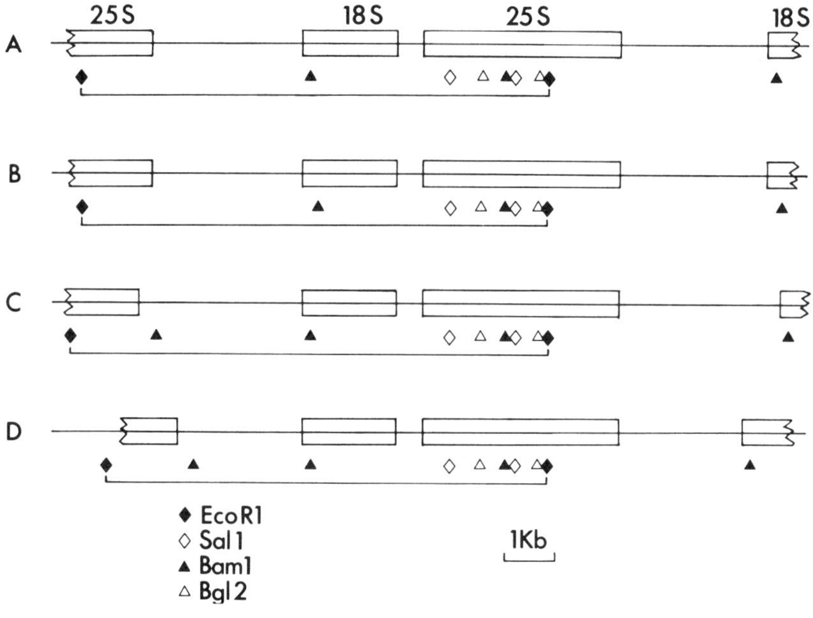

Fig. 10. Physical maps of the rDNA repeats of barley and wheat using restriction endonuclease targets as markers. (A) map for the major wheat rDNA repeat, (B) map for the minor wheat rDNA repeat, (C) map for the longer barley rDNA repeat, (D) map for shorter barley rDNA repeat.

internal to the coding sequence is identical for both classes of wheat rDNA. The Bam HI•Eco RI fragment that extends from the 18S coding sequence into the nontranscribed spacer sequence is 0.15 Kb longer in the minor species illustrated in Fig. 10B. The larger barley repeat (Fig. 10C) is 9.93 Kb and the smaller repeat D is 8.95 Kb. The difference in length of the two barley repeats can be attributed to differences in the lengths of the nontranscribed spacer in the two length classes of barley rDNA repeat. DNA fragments produced by restriction endonucleases and internal to the coding portion of the genes are identical in both types of barley repeat.

In summary, all the detected length variation between various rDNA repeats of barley and wheat can be attributed to variation in the length of the nontranscribed spacer. Experiments in which purified nontranscribed spacer have been hybridized back to total wheat DNA fractionated on gels suggest that the nontranscribed spacer sequence is present only in the context of the ribosomal DNA repeat and not present in other chromosomal locations. This conclusion is supported by *in situ* hybridization of rDNA probes to metaphase chromosomes (Fig. 11). Tritium-labeled cRNA prepared by using cloned rDNA as template was hybridized to wheat chromosomes *in situ*. Fig. 11 shows that the probe hybridizes only the major sites on chromosomes 1B and 6B known to be rDNA loci (Flavell and O'Dell, 1976; Liang et al., 1977; Appels et al., 1979).

Conclusions

Cereal DNA prepared from nuclei on CsCl-EtBr density gradients can be cloned with an efficiency slightly lower than similarly prepared *E. coli* DNA. The efficiency is high enough to envisage total genome cloning of plant DNA. DNA clones characterized so far are stable in *E. coli*. Insert length in clones has not been studied systmatically. However, clones obtained for several repeated sequences and genes contain inserts of the length expected from restriction digests of genomic DNA.

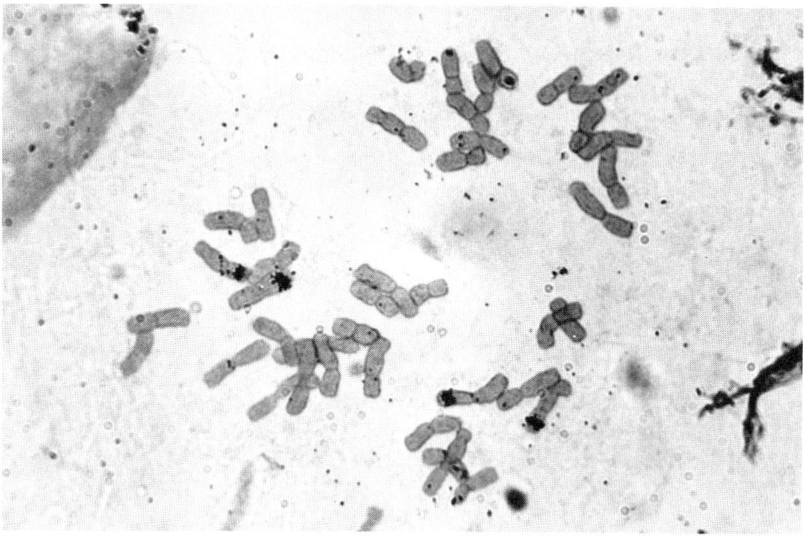

Fig. 11. *In situ* hybridization to wheat chromosomes of cRNA prepared from molecular clones of wheat ribosomal DNA. ^{3}H-labeled cRNA prepared from a plasmid containing the wheat EcoR I repeat of the ribosomal DNA was hybridized to wheat chromosomes and exposed to photographic emulsion as described by Appels et al. (1979).

References

Appels, R.; Gerlach, W. L.; Dennis, E. S.; Swift, H.; and Peacock, W. J. (1979) Organisation of ribosomal genes in cereals. (Manuscript in preparation.)

Bedbrook, J. R.; Kolodner, R.; and Bogorad, L. (1977) *Zea mays* chloroplast ribosomal RNA genes are part of a 22,000 base pair inverted repeat. *Cell 11*, 739-49.

Bedbrook, J. R.; Coen, D. M.; Beaton, A. R.; Bogorad, L.; and Rich, A. (1979a) Location of the single gene for the large subunit of ribulose bisphosphate carboxylase on the maize chloroplast chromosome. *J. Biol. Chem. 254*, 905-10.

Bedbrook, J. R.; Jones, J.; Thompson, R.; and Flavell, R. (1979b) On the molecular properties of the telomeres of rye chromosomes. (Manuscript in preparation.)

Bennett, M. D.; Gustafson, J. P.; and Smith, J. B. (1977) Variation in nuclear DNA in the genus *Secale. Chromosoma (Berline) 61*, 149-76.

Bird, A. P. (1978) Use of restriction enzymes to study eukaryotic DNA methylation. II: The symmetry of methylated sites supports semi-conservative copying of the methylation pattern. *J. Mol. Biol. 118*, 49-60.

Bird, A. P., and Southern, E. M. (1978) Use of restriction enzymes to study eukaryotic DNA methylation. I: The methylation pattern in ribosomal DNA from *Xenopus laevis. J. Mol. Biol. 118*, 27-47.

Brutlag, D.; Fry, K.; Nelson, T.; and Hung, P. (1977) Synthesis of hybrid bacterial plasmids containing highly repeated satellite DNA. *Cell 10*, 509.

Coen, D. M.; Bedbrook, J. R.; Bogorad, L.; and Rich, A. (1977) Maize chloroplast DNA fragment encoding the large subunit of ribulose bisphosphate carboxylase. *Proc. Natl. Acad. Sci., U.S.A. 74*, 5487-91.

Flavell, R. B., and O'Dell, M. (1976) Ribosomal RNA genes on homoeologous chromosomes of groups 5 and 6 in hexaploid wheat. *Heredity 37*, 377-85.

Flavell, R. B.; Bennett, M. D.; Smith, J. B.; and Smith, D. B. (1974) Genome size and the proportion of repeated nucleotide sequence DNA in plants. *Biochem. Genet. 12*, 257-69.

Flavell, R. B.; O'Dell, M.; and Jones, J. (1980) Cereal genome studies and plant breeding research. This volume.

Gerlach, W. L., and Bedbrook, J. R. (1979) (Manuscript submitted.)

Grunstein, M., and Hogness, D. S. (1975) Colony hybridization: A method for the isolation of cloned DNAs that contain a specific gene. *Proc. Nat. Acad. Sci., U.S.A. 72*, 3961-65.

Hamer, D. H., and Thomas, C. A. (1975) The cleavage of *Drosophila melanogaster* DNA by restriction endonucleases. *Chromosoma 49*, 243-67.

Hattman, S.; Keister, T.; and Gottehrer, A. (1978) Sequence specificity of DNA methylases from *Bacillus amyloliquefaciens* and *Bacillus brevis. J. Mol. Biol. 124*, 701-11.

Hemleben, V., and Grierson, D. (1978) Evidence that in higher plants the 25S and 18S rRNA genes are not interspersed with genes for 5S rRNA. *Chromosoma 65*, 353-58.

Liang, G. H.; Wang, A. S.; and Phillips, R. L. (1977) Control of ribosomal RNA gene multiplicity in wheat. *Can. J. Genet. Cytol. 19*, 425-35.

Maniatis, T.; Hardison, R. C.; Lacy, E.; Lauer, J.; O'Connell, C.; and Quon, D. (1978) The isolation of structural gene libraries of eukaryotic DNA. *Cell 15*, 687-701.

Mann, M. B., and Smith, H. O. (1977) Specificity of Hpa II and Hae III DNA methylases. *Nucleic Acids Res. 4*, 4211-21.

Peacock, W. J.; Lohe, A. R.; Gerlach, W. L.; Dunsmuir, P.; Dennis, E. S.; and Appels, R. (1978) Fine structure and evolution of DNA in heterochromatin. In *Cold Spring Harbor Symp. Quant. Biol.* pp. 1121-35.

Rimpau, J.; Smith, D.; and Flavell, R. (1978) Sequence organisation analysis of the wheat and rye genomes by interspecies DNA/DNA hybridization. *J. Mol. Biol. 123*, 327-59.

Characteristics of T-DNA in Crown Gall Tumors

M.-D. Chilton, J. McPherson, R. K. Saiki, M. F. Thomashow, R. C. Nutter, S. B. Gelvin, A. L. Montoya, D. J. Merlo, F.-M. Yang, D. J. Garfinkel, E. W. Nester, and M. P. Gordon

The plant cancer crown gall is incited by inoculation of wound sites with *Agrobacterium tumefaciens*. The resulting gall contains neoplastic cells that possess altered growth characteristics and metabolism. Tumor cells can grow *in vitro* in the absence of the inciting bacterium, and, unlike normal plant callus, the tumor tissue grows luxuriantly without exogenously supplied phytohormones (White and Braun, 1942).

Tumor cells *in vitro* synthesize novel amino-acid derivatives called *opines*. The most extensively characterized opines are those of the octopine group—octopine and octopinic acid (Ménagé and Morel, 1964), lysopine (Lioret, 1957; Biemann, et al., 1960), and histopine (Kemp, 1977)—and those of the nopaline group—nopaline

Note: This research was supported by Grant Number CA13015 from the National Cancer Institute and Grant Number NP194 from the American Cancer Society. R. C. Nutter was supported by fellowhip # CA06182 from the National Cancer Institute. M. F. Thomashow and S. B. Gelvin were supported by fellowships #DRG-209F and #DRG-213F, respectively, from the Damon Runyan-Walter Winchell Cancer Fund. We thank Dr. Armin C. Braun for providing us with plant tissue cultures. We are indebted to Dr. Braun, Dr. Fred Meins, and Dr. Howard Bonnett for illuminating discussions. We thank Dr. Francis Quétier for guidance in the isolation of plastid DNA from plant tissues grown *in vitro*. We are indebted to Dr. Allen Kerr for the bacterial strains K305 and K309.

(Ménagé and Morel, 1965) and ornaline (Firmin and Fenwick, 1977). The structures of these compounds are illustrated in Fig. 1. Recently, a new opine has been identified in octopine-type tumors and designated *agropine* (Firmin and Fenwick, 1978); its chemical structure is not yet known. The known opines are derived by one enzymatic step from the normal L-amino acids. The octopine family of opines is produced by a synthase that has been purified to homogeneity (Hack and Kemp, 1977; Otten et al., 1977; Kemp et al., 1978). Similarly, the nopaline family of opines is produced by a homogeneous synthase (J. D. Kemp, personal communication). Opines are not found in normal plant tissue (Gordon et al., 1979; Holderbach and Beiderbeck, 1976; Kemp, 1976). The class of

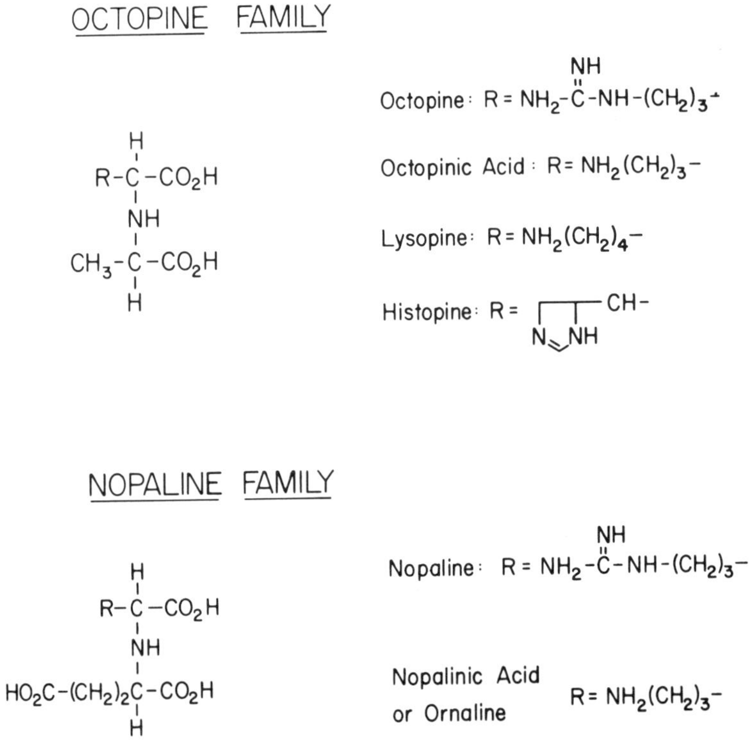

Fig. 1. Structures of opines synthesized by crown gall tumors (see text).

opines produced by the tumor is determined by the inciting bacterial strain (Petit et al., 1970). In general, the bacterium possesses specific inducible uptake and catabolic enzymes that allow it to utilize the class of opines produced by the tumors it incites (Petit et al., 1970; Petit and Tempé, 1978). The specific production of opines in response to specific pathogenic Agrobacterium strains led Morel and his collaborators to suggest that bacterial genes transferred to the plant cell may be responsible for synthesis of opines (Petit et al., 1970). The synthetic enzyme expressed in the plant tumor is clearly not identical to the enzyme for opine utilization expressed in Agrobacterium. Mutant bacteria that have lost the ability to utilize opines still incite tumors that produce opines in the expected amounts (Klapwijk et al., 1976; Montoya et al., 1977; Petit and Tempé, 1978). Recently, genetic-mapping evidence has borne out the conclusion that these are distinct genetic loci (Schell, 1978).

Giant Plasmids

Large tumor-inducing (Ti) plasmids ($100 - 150 \times 10^6$ daltons) in *Agrobacterium tumefaciens* (Zaenen et al., 1974) carry genetic information required for oncogenicity (Van Larebeke et al., 1974; Watson et al., 1975). This has been demonstrated for two Ti plasmids by heat curing (loss of oncogenicity correlates with loss of the Ti plasmid) and for numerous Ti plasmids by genetic manipulation. The Ti plasmids are conjugative, and their transfer (Kerr, 1969; Kerr, 1971) was first discovered by transfer of the oncogenicity trait from one Agrobacterium strain to another *in planta*. This was later shown to be mediated by Ti-plasmid transfer (Van Larebeke et al., 1975; Watson et al., 1975). Ti-plasmid genes allow Agrobacterium to catabolize specific opines (Van Larebeke et al., 1974; Watson, et al., 1975). This fact afforded the first-known selectable genetic trait coded by Ti plasmids and made it possible to detect Ti-plasmid transfer *in vitro*. The conjugative plasmid RP4 was found to mobilize the Ti plasmids (Chilton et al., 1976; Van Larebeke et al., 1977). A method was devised for transforming Agrobacterium with purified Ti-plasmid DNA (Holsters et al.,

1978). Further, the Ti plasmids were found to conjugate *in vitro* when induced with a high concentration of the specific opine whose catabolism they coded (Kerr et al., 1977; Genetello et al., 1977; Petit et al., 1978). In all of these plasmid-transfer studies, the Ti plasmids were found to be associated with oncogenicity. Transfer of opine catabolism (the selected trait) was always associated with transfer of the Ti plasmid and oncogenicity; transfer of oncogenicity (the screened trait) *in planta* was associated with transfer of opine catabolism and the Ti plasmid. Loss of oncogenicity (curing experiments) was associated with loss of opine catabolism and the Ti plasmid.

The location on Ti plasmids of genes for opine catabolism is reasonable from an evolutionary point of view. Opines appear to play a central role in crown gall biology. One can argue that they provide a selective advantage to the pathogen by two means. They provide positive nutritional selection and they induce the spread of the Ti plasmid by creating new pathogenic strains. From future studies of opine traits and DNA homology for diverse Ti plasmids, we may be able to trace the evolutionary steps by which the wide-host-range Ti plasmids emerged.

Wide-host-range Ti plasmids are indeed diverse, as can be discerned from DNA-hybridization studies (Currier and Nester, 1976; Drummond and Chilton, 1978). There is as little as 20% homology between the most distantly related plasmids; Tm measurements indicate that the homologous sequences are extensively mismatched (Drummond and Chilton, 1978). The Sma I cleavage patterns of Ti plasmids are diverse (Fig. 2) (Sciaky et al., 1978), although the wide-host-range octopine-type Ti plasmids are a homogeneous group. The octopine-utilization trait is carried on at least one non-tumor-inducing plasmid (Sciaky et al., 1978) that bears little homology to the octopine Ti plasmid (Drummond and Chilton, 1978). Ti plasmids from two Greek and two Australian strains oncogenic on grape vines carry the octopine trait and have Sma I cleavage patterns that bear no resemblance to that of the canonical octopine Ti plasmid (Fig. 3). The extent of the repertory of Ti-plasmid types in nature is not yet clear.

Recently, Dénarié and his collaborators discovered that even the "cured" strains of Agrobacterium, thought to be plasmidless,

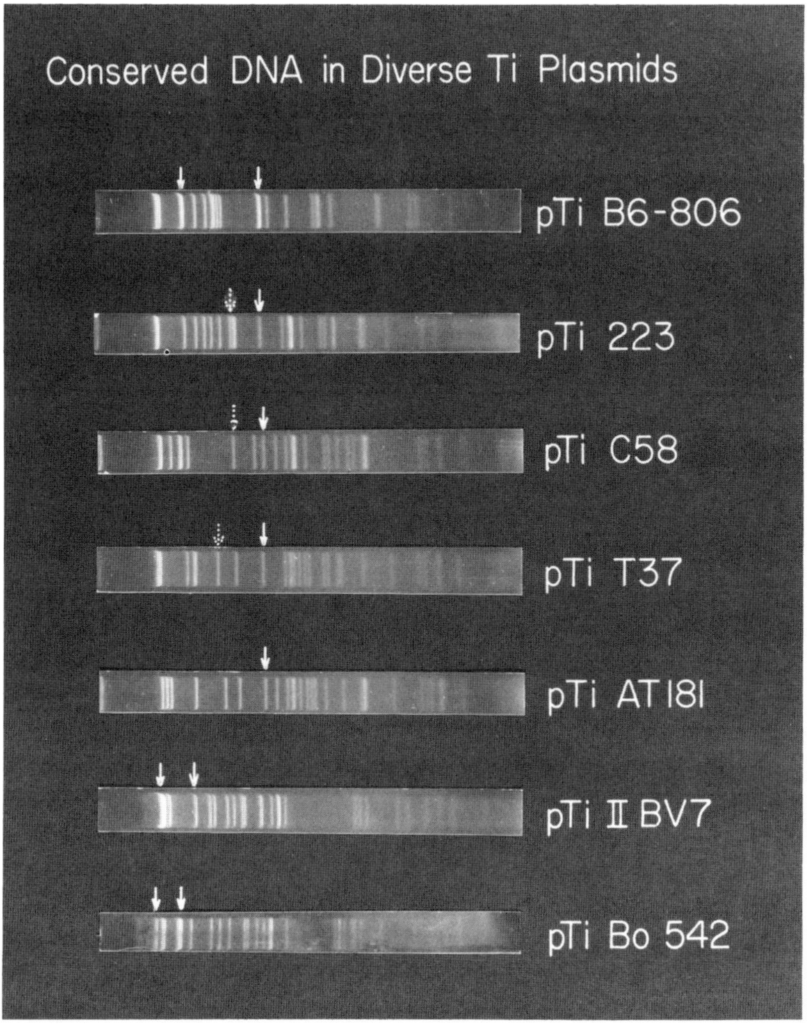

Fig. 2. Sma I fingerprints of Ti plasmids. The conditions of digestion and agarose-gel electrophoresis are described in detail by Sciaky et al. (1978). Arrows mark the positions at which Hind III fragment 1 of pTi B6-806 hybridizes well (solid arrows) or poorly (dotted arrows) with each Ti plasmid (see Fig. 5 and text) (Chilton et al., 1978c; DePicker et al., 1978).

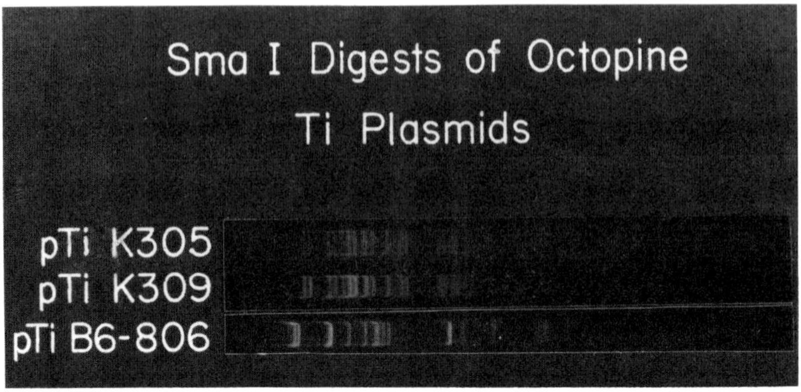

Fig. 3. Sma I fingerprints of octopine Ti plasmids found in two Australian grape gall pathogens, K305 and K309, isolated by Dr. Allen Kerr. The cleavage pattern of the common wide-host-range octopine Ti plasmid pTi B6-806 is shown for comparison. Octopine-type Ti plasmids are a rarity in Australia. Strains K305 and K309 were isolated on the same property and appear identical. These plasmids carry the octopine trait in that pure plasmid DNA produces octopine-utilizing transformants of strain A136 (a strain unable to utilize octopine or nopaline).

contain giant cryptic plasmids (Casse et al., 1979). There is at present no reason to suppose that these cryptic plasmids play a role in tumor induction. The giant cryptic plasmids of three avirulent Agrobacterium strains examined are diverse; the Sma I cleavage patterns of the three are strikingly different (Fig. 4). Nevertheless, until a truly plasmidless Agrobacterium strain is available for use as a recepient in Ti-plasmid-transfer studies, we cannot rule out the possibility that these giant cryptic plasmids may interact with the known Ti plasmids to bring about tumor induction.

Ti-Plasmid DNA in Octopine-Type Tumor Cells

Crown gall tumor cells appear to be permanently altered by their transient contact with *Agrobacterium tumefaciens.* The specific

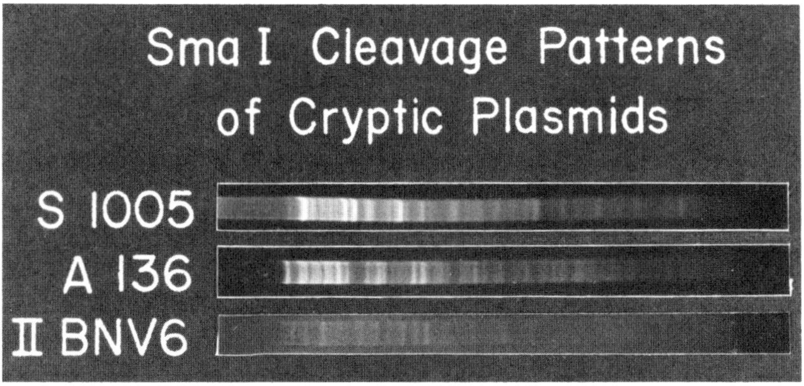

Fig. 4. Sma I Fingerprints of giant cryptic plasmids isolated from avirulent Agrobacterium strains S1005, A136, and IIBNV6. Details of digestion and electrophoretic separation are described by Sciaky et al. (1978).

synthesis of opines is dictated by the bacterium, not by the plant. This led Morel and his collaborators to suggest that the tumor cell contains bacterial genes (Petit et al., 1970). Since opine-synthesis determinants are now known to be Ti-plasmid-borne traits, Morel's model predicts the presence of Ti-plasmid DNA in crown gall tumor cells.

This possibility was tested soon after the Ti plasmid was discovered. Using "driver-DNA reactions," with labeled Ti-plasmid DNA as probe, we measured the rate of reassociation of a trace of Ti-plasmid DNA alone or in the presence of a high concentration of tumor DNA. The tumor DNA was isolated from axenic crown gall tumor tissue, and rigorous precautions were taken to rule ot out accidental contamination of the DNA preparation with bacterial DNA from other sources. If tumor DNA contained copies of the Ti plasmid, addition of this driver to the Ti-plasmid-probe reaction should accelerate probe renaturation in a predictable way. Reconstruction reactions showed that as little as one copy of the entire Ti plasmid per diploid plant cell could be detected by this technique. When tumor DNA from a cloned tumor line was examined, we were unable to detect the presence of the entire Ti plasmid (Chilton et al., 1975; Gordon et al., 1979).

Detection of T-DNA in Octopine Tumors

Although the driver-DNA technique is very powerful, one of its fundamental requirements is that the correct probe be used. If one used the entire Ti plasmid as probe but only 5% of it were in the tumor, only 5% of the labeled probe DNA would renaturate at an increased rate. This would be very difficult to detect experimentally. Because the biological characteristics of crown gall tumor cells so strongly suggested Ti-plasmid gene transfer, it appeared worthwhile, even in the face of the negative results described above, to look for a small part of the Ti plasmid in tumor DNA. For this purpose, we separated the Ti plasmid into specific fragments and used each fragment as a probe in further driver-DNA experiments. Since we had no clue as to what part of the plasmid was likely to be in the tumor cell, we were obliged to test every fragment in a "brute-force" series of DNA-renaturation kinetic experiments.

An octopine Ti plasmid (pTi B6-806) was fragmented by Sma I into 19 visible bands (Fig. 2), and each was used as probe in a tumor-DNA-driven reaction. Only fragments 3b and 10c gave positive results (Chilton et al., 1977; Chilton et al., 1978a). Mapping of the Sma I digest fragments of pTi B6-806 revealed that 3b and 10c are contiguous on the plasmid (Chilton et al., 1978b). Part of the fragment map is reproduced in Fig. 5. The part of the Ti plasmid detected in tumors is called T-DNA. The T-DNA of tumor line E9 appeared to be present in multiple copies (Chilton et al., 1977). In a detailed analysis of copy number, we found (unpublished results) that the right edge of the T-DNA in E9 tumor is present in high copy while the left edge is present in about single copy, that is, the various copies of T-DNA in this cloned tumor line are clearly not all of equal length. The multiple copies of the right portion (*left* and *right* are arbitrarily defined by the map orientation of Fig. 5) seemed at first to suggest that this was the essential region of the T-DNA and that perhaps the shorter copies had arisen by deletion of the DNA not under positive selection in the tumor cell. Subsequent studies of two other octopine-type tumor lines soon showed that this was not the case. One of these tumor lines turned out to have none of the right-hand end of the T-DNA and only one of two copies of the left-hand portion

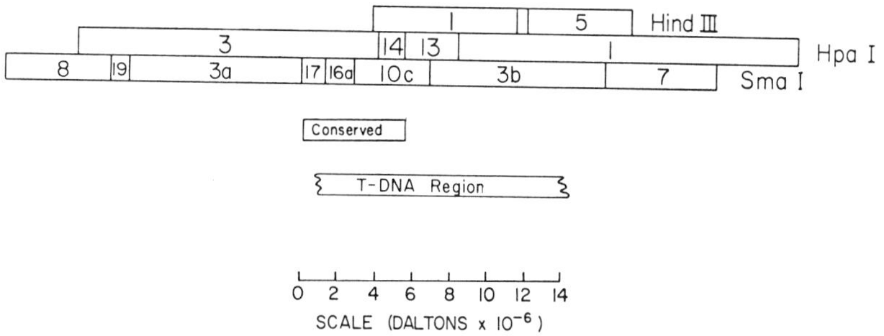

Fig. 5. Portion of the physical map of pTi B6-806 containing fragments detected in crown gall tumors ("T-DNA region"). Highly conserved DNA (Chilton et al., 1978c) overlaps T-DNA (see text).

(unpublished results). The essential part of T-DNA is therefore the left end, the part that maps under Hind III fragment 1 (Fig. 5). The presence of multiple short copies of the nonessential portion in the E9 tumor line remains an intriguing puzzle.

In an effort to determine whether the essential part of the T-DNA common to all octopine tumors studied is also common to all Ti plasmids, a labeled probe DNA containing this region (Hind III fragment 1) was hybridized to Southern (1975) blots of Sma I cleavage products of a variety of Ti plasmids (Fig. 2). This probe hybridized to every Ti plasmid examined and, indeed, to a fragment the size of octopine-plasmid-digest band 10c in most cases (Fig. 2) (Chilton et al., 1978c; DePicker et al., 1978). A detailed hybridization study using six contiguous labeled probe fragments delineated the exact boundaries of a highly conserved DNA element found on every wide host range Ti plasmid examined (Fig. 5) (Chilton et al., 1978c). DePicker and her associates (1978) showed that this part of the Ti plasmid codes for a function essential to oncogenicity: RP4 insertion (cointegration) within this conserved part of the Ti plasmid abolishes the ability of the bacterium to incite tumors. We know from recent analysis of several octopine-type tumors by Southern blotting analysis (unpublished results) that T-DNA includes much of the conserved DNA region.

Transcription of T-DNA in Octopine-Type Tumor Cells

^{32}P-pulse-labeled RNA isolated from octopine-type tumor cells hybridizes with Southern (1975) blots of Ti-plasmid fragments (Drummond et al., 1977). Transcripts of Sma I fragment 3b were detected, but not of fragment 10c (the conserved DNA region). Other octopine-type tumors subsequently examined (and, indeed, the same one at a very low level) were found to contain fragment-10c transcripts (Lederboer, 1978; Gurley et al., 1979). T-DNA transcripts were found in the polyadenylated fraction of tumor-cell RNA (Lederboer, 1978). Indeed, the polyadenylated fraction of polysomal RNA contains T-DNA transcripts (Fig. 6). There seems little reason to doubt that T-DNA genetic information functions in the plant cell and that specific T-DNA-encoded proteins will be identified in the tumor. A prime candidate for one such protein would be the enzyme octopine synthase.

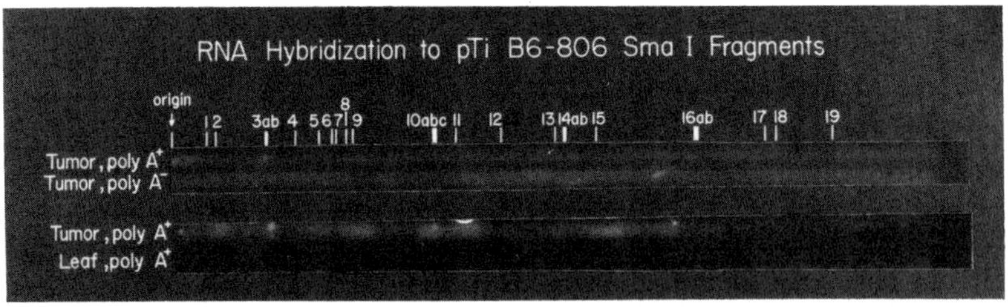

Fig. 6. Hybridization of polysomal RNA isolated from crown gall tumor tissue to Southern blots of Ti-plasmid Sma I-digest fragments. The pattern of DNA fragments used to prepare the Southern blots is presented as a line drawing. RNA was preapred from E1 tumor callus tissue and from normal leaf by a modification of the method of Palmiter (1974). Polyadenylated RNA was isolated by chromatography using oligo-dT cellulose. Both the polyadenylated (A+) and nonpolyadenylated (A−) fractions (Aviv and Leder, 1972) from tumor and normal tissue were end-labeled *in vitro* (Doel et al., 1977) by γ-^{32}P-adenosine triphosphate (New England Nuclear, 3000 Ci/mMol). Labeled RNA (2×10^6 cpm/ml) was hybridized with Southern (1975) blots in 2 × SSC, 0.1% sodium dodecyl sulfate at 67°C), for 20 hours. After washing (2 × SSC, 67 °C), strips were autoradiographed at −70 °C with intensification screens for 5 days.

Ti-Plasmid DNA in Nopaline-Type Tumor Cells

The common DNA region is preserved on all of the wide-host-range Ti plasmids examined. But would this common DNA form a part of the T-DNA of nopaline-type tumor lines? To answer this question, a cloned T37 tumor line from Dr. Armin Braun's collection was analysed by using the driver-DNA approach outlined above (unpublished results). T-DNA of this tumor line turned out to encompass completely the highly conserved DNA region shown in Fig. 5 and to include 18×10^6 daltons of the Ti plasmid.

This nopaline tumor line is of unusual interest because it has been used in a curing experiment to produce plants free from the tumorous characteristics (Braun and Wood, 1976; Turgeon et al., 1976). The cloned parental tumor line BT37, of Havana-tobacco origin, was grafted to the apexes of decapitated Turkish tobacco plants. Rarely, a shoot emerged from the grafted tissue. The shoot gradually developed a normal morphology and flowers that set seed. The F1 progeny of the plants of this type were found to be completely normal in every respect, including their susceptibility to crown gall. The original graft shoot, although of largely normal morphology, retained two tumorous characteristics: (1) various tissues were found to synthesize nopaline, the opine characteristic of the parental tumor line, and (2) these same tissues (leaf, flower petal, filament) were found to revert to tumorous growth habit *in vitro*. Postmeiotic tissues, in contrast, were free of these tumorous traits. Callus tissue derived from haploid plants of anther origin was normal and nopaline-free. F1 progeny derived from either male or female gametes or both of the teratoma plant were uniformly normal and nopaline-free. A DNA hybridization analysis of the tumorous and normal tissue lines derived from a teratoma plant (unpublished results) revealed that T-DNA was detectable only in the nopaline-containing tissues that behaved as tumors *in vitro*. Postmeiotic tissues were found to be free of T-DNA. These results suggested that the mechanism for reversal of the tumorous state involved excision of T-DNA at meiosis.

However, that complete reversal (and, presumably, elimination of T-DNA) is possible as a rare event without intervention of meiotic events has been indicated by finding a completely normal grafted shoot derived from a cloned teratoma (Turgeon et al., 1976). DNA from this revertant plant tissue is being tested for the presence of T-DNA. Similar regeneration experiments with octopine-type and "null-type" tumor clones have been reported by Sacristán and Melchers (1977). Reversion is achieved *in vitro* without grafting. The recovered plant tissues have not been analyzed for T-DNA content.

From the rare vegetative recovery of BT37 tumor tissue, it is clear that meiosis is not required for the loss of the tumorous phenotype; normal cells can simply segregate or develop from the cloned tumor line. We presume that the normal-appearing tissue obtained in this way will turn out to be free from T-DNA for there is a complete correlation thus far between the presence of T-DNA and the expression of the nopaline trait in the experiments with the teratoma plant developed by grafting. The absence of T-DNA from postmeiotic tissues of the teratoma plant may be ascribed to one of two explanations: (1) The teratoma plant may be a mosaic of tumorous and normal cells, with only the latter yielding viable meiotic products. This model implies that the mechanism of curing is the spontaneous segregation of normal cells from the cloned tumor line as indicated by the vegetative-recovery phenomenon described above. (2) There may be a unique mechanism for eliminating T-DNA at meiosis. The occasional segregation of normal cells from the cloned tumor line may be an entirely different type of reversal.

Where Is the T-DNA?

The studies of BT37 tumor cells and their ability to lose (excise or segregate?) the two to four copies of T-DNA they contain raises the issue of the location and attachment site, if any, of T-DNA in the plant cell. The seemingly frequent and complete loss of T-DNA does not seem harmonious with the view that it is inserted into chromosomal DNA.

To look directly for T-DNA in chloroplast and mitochondrial DNA, we have isolated these DNA fractions from purified organelles (Vedel et al., 1976; Quétier and Vedel, 1977). The plastid-

DNA fractions appear free from chromosomal DNA. Cleavage with the restriction endonuclease Bst I gives a simple pattern of fragments as detected by electrophoresis in agarose (Fig. 7); however, the "chloroplast" DNA fraction is significantly contaminated with mitochondrial DNA. A Southern blot prepared from these fragments was hybridized with labeled pTi T37 plasmid DNA with negative results; we conclude that the T-DNA is not located in either of these plastid fractions. By this technique, we calculate that we should easily be able to detect T-DNA fragments. If all T-DNA were in the organellar DNA and if that DNA made up 1% to 5% of total tumor DNA, then the T-DNA should be enriched 20 to 100 times in the plastid DNA compared to the total tumor DNA. Because T-DNA fragments are indeed detectable in total tumor DNA by this technique (unpublished results), the preliminary negative data appear significant.

If T-DNA is not in mitochondrial or chloroplast DNA and if it segregates with a frequency higher than expected for chromosomal DNA, can one imagine any alternative location of T-DNA?

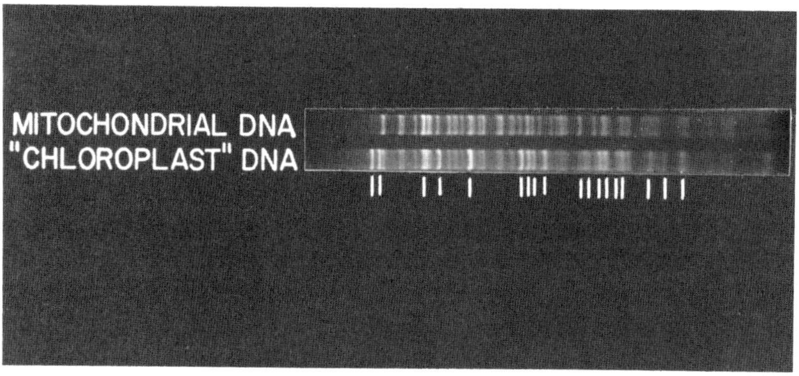

Fig. 7. Bst I cleavage of BT37 tobacco tumor chloroplast and mitochondrial DNA fractions. One microgram of each DNA sample was digested to completion with Bst I (a gift of Dr. Rich Meagher). This enzyme is an isoschizomer of Bam HI. The "chloroplast" DNA digest contains mitochondrial DNA digest fragments; the line drawing below the photo indicates the bands we assign to chloroplast DNA.

It is conceivable that T-DNA is attached to some as yet unknown free replicon in the plant cell. Genetic characteristics of controlling elements in maize (Nevers and Saedler, 1977) as well as recent developments in Texas male sterile maize genetics (R. Mans, personal communication) suggest that there are mobile DNA elements in the plant genome whose behavior is as yet ill defined. The location of T-DNA in the plant cell will be elucidated from molecular cloning experiments, which should allow isolation of T-DNA from the plant genome together with the plant DNA (if any) to which it is attached.

Prospects for Modification of Plants

The Ti plasmid is a potential vector for artificial introduction of novel DNA into plants. The host range of crown gall disease is very wide and includes many kinds of dicotyledonous plants. If a protoplast-transformation system can be developed, the host range of Ti plasmid DNA may be extended further. Limitation of host range may be simply ascribable to the inability of *A. tumefaciens* to bind to cell walls of monocots (Lippincott et al., 1977). The Ti-plasmid DNA is very stably maintained by the transformed tissue. It appears to be attached to some other kind of DNA, presumably plant DNA. Available evidence, although not definitive, is consistent with the view that the T-DNA element is in nuclear DNA. It is in part actively transcribed, and the transcripts are polyadenylated and appear in the polysomal fraction. It seems very likely that proteins are coded by these transcripts, including the enzymes that are responsible for opine synthesis.

There are two technical barriers that remain. Indeed, these may have a common solution: the T-DNA is apparently not seed transmitted and the transformed plant does not appear morphologically normal. If we can achieve an understanding of the molecular basis of the transformation process and if it is not identical to the foreign-DNA-insertion process itself, we may eventually learn to switch the transforming genes off at will and restore the altered plant cell to normal growth characteristics.

The direct exploitation of Ti plasmids in Agrobacterium as vectors for introducing novel DNA into plants is impeded by the restrictions on recombinant-DNA manipulation posed by the N.I.H. guidelines. Two aspects of the approach are prohibited when eukaryotic (plant or animal) DNA is chosen as the novel DNA to be transferred. Higher plant DNA cloning, for example, may be performed at P2EK1 or P1EK2 level of containment. This means that (1) plant-DNA fragments to be cloned must be ligated into nonconjugative *E. coli* plasmids (not the Ti plasmid) and (2) the recombinant DNA must be maintained in *E. coli K12* (not Agrobacterium). With permission, an HV1 system could be used to harbor such recombinant DNA; however, Agrobacterium fails to qualify for HV1 certification because it is a pathogen, and indeed its very pathogenicity is an essential feature of this approach to plant modification. Special permission from the Recombinant Advisory Council will be required before Agrobacterium and the Ti plasmid may be used for introduction of desirable eukaryotic DNA into crop plants. The alternative of transforming plant protoplasts directly with recombinant DNA is allowed by the N.I.H. guidelines at P1 containment. However, the technical feasibility of this approach for the plant one wishes to use as host is not assured.

Conclusions

One can clearly discern the areas in which crown gall research needs to advance. From molecular cloning experiments currently underway in several laboratories, T-DNA will be reisolated from the from the tumor genome, together with the contiguous DNA segments. Studies of such clones will reveal whether T-DNA is indeed attached to plant DNA and, if so, to what kind of plant DNA. Sequencing studies may reveal at the molecular level the details of the mechanism by which T-DNA is joined (presumably) to plant DNA. In addition, detailed studies of the transcription of T-DNA in the tumor cell should show whether T-DNA possesses promoters active in the plant cell. The results will bear on the interesting possibility that either the T-DNA or the contiguous plant DNA may be transcribed by readthrough. The mechanism by which Ti-plasmid DNA alters the plant cell to its own ends will show us at the molecular level the details of a natural case of genetic engineering by a prokaryotic organism. It is a lesson in strategy that is well worth the attention of eukaryotic organisms with similar intentions.

References

Aviv, H., and Leder, P. (1972) Purification of biologically active globin messenger RNA by chromatography on oligothymidylic acid-cellulose. *Proc. Nat. Acad. Sci., U.S.A. 69*, 1408-12.

Biemann, K.; Lioret, C.; Asselineau, J.; Lederer, E.; and Polonski, J. (1960) Sur la structure chimique de la lysopine, nouvel acide aminé isolé des tissues de crown-gall. *Bull. Soc. Chim. 17*, 979-91.

Braun, A. C., and Wood, H. N. (1976) Suppression of the neoplastic state with the acquisition of specialized functions in cells, tissues and organs of crown gall teratomas of tobacco. *Proc. Nat. Acad. Sci., U.S.A. 73*, 496-500.

Casse, F.; Boucher, C.; Julliot, J.S.; Michel, M.; and Dénarié, J. (1979) Identification and characterization of large plasmids in *Rhizobium meliloti* using agarose gel electrophoresis. *J. Gen. Microbiol. 113*, 229-42.

Chilton, M. -D.; Farrand, S. K.; Eden, F. C.; Currier, T. C.; Bendich, A. J.; Gordon, M. P.; and Nester, E. W. (1975) Is there foreign DNA in crown gall tumor DNA? In *Modification of the Information Content of Plant Cells* (R. Markham, D. R. Davies, D. A. Hopwood, and R. W. Horne, eds.). Elsevier, New York.

Chilton, M. -D.; Farrand, S. K.; Levin, R.; and Nester, E. W. (1976) RP4 promotion of transfer of a large Agrobacterium plasmid which confers virulence. *Genetics 83*, 609-18.

Chilton, M. -D.; Drummond, M. H.; Merlo, D. J.; Montoya, A. L.; Sciaky, D.; Gordon, M. P.; and Nester, E. W. (1977) Stable incorporation of plasmid DNA into higher plant cells: The molecular basis of crown gall tumorigenesis. *Cell 11*, 263-71.

Chilton, M. -D.; Drummond, M. H.; Gordon, M. P.; Merlo, D. J.; Montoya, A. L.; Sciaky, D.; Nutter, R.; and Nester, E. W. (1978a) Foreign genes of plasmid origin detected in crown gall tumor. *Microbiology 1978*, 136-38.

Chilton, M. -D.; Montoya, A. L.; Merlo, D. J.; Drummond, M. H.; Nutter, R.; Gordon, M. P.; and Nester, E. W. (1978b) Restriction endonuclease mapping of a plasmid that confers oncogenicity upon *Agrobacterium tumefaciens* strain B6-806. *Plasmid 1*, 254-69.

Chilton, M. -D.; Drummond, M. H.; Merlo, D. J.; and Sciaky, D. (1978c) Highly conserved DNA of Ti plasmids overlaps T-DNA maintained in plant tumors. *Nature 275*, 147-49.

Currier, T. C., and Nester, E. W. (1976) Evidence for diverse types of large plasmids in tumor-induing strains of *Agrobacterium. J. Bacteriol. 126*, 157-65.

DePicker, A.; Van Montagu, M.; and Schell, J. (1978) Homologous DNA sequences in different Ti-plasmids are essential for oncogenicity. *Nature 275*, 150-52.

Doel, M. T.; Houghton, M.; Cook, E. A.; and Carey, N. H. (1977) The presence of ovalbumin m-RNA coding sequences in multiple restriction fragments of chicken DNA. *Nucleic Acids Res. 4*, 3701-13.

Drummond, M. H., and Chilton, M. -D. (1978) Agrobacterium Ti plasmids share extensive regions of DNA homology. *J. Bacteriol. 136*, 1178-83.

Drummond, M. H.; Gordon, M. P.; Nester, E. W.; and Chilton, M. -D. (1977) Foreign DNA of bacterial plasmid origin is transcribed in crown gall tumors. *Nature 269*, 535-36.

Firmin, J. L., and Fenwick, R. G. (1977) N2 (1,3-dicarboxypropyl) ornithine in crown gall tumors. *Phytochemistry 16*, 761-62.

Firmin, J. L., and Fenwick, G. R. (1978) Agropine: A major new plasmid-determined metabolite in crown gall tumors. *Nature 276*, 842-44.

Genetello, C.; Van Larebeke, N.; Holsters, M.; DePicker, A.; Van Montagu, M.; and Schell, J. (1977) Ti plasmids of Agrobacterium as conjugative plasmids. *Nature 265*, 561-63.

Gordon, M. P.; Farrand, S. K., Sciaky, D.; Montoya, A. L.; Chilton, M. -D.; Merlo, D. J.; and Nester, E. W. (1979) The crown gall problem. In *A Symposium on the Molecular Biology of Plants* (I. Rubenstein, ed.). Academic Press, New York, pp. 291-313.

Gurley, W. B.; Kemp, J. D.; Albert, M. J.; Sutton, D. W.; and Callis, J. (1979) Transcription of Ti plasmid derived sequences in three octopine-type crown gall tumor lines. *Proc. Nat. Acad. Sci., U.S.A. 76*, 2828-32.

Hack, E., and Kemp, J. D. (1977) Comparison of octopine, histopine, lysopine and octopinic acid synthesizing activities in sunflower crown gall tissues. *Biochem. Biophys. Res. Commun. 78*, 785-91.

Holderbach, E., and Beiderbeck, R. (1976) Octopingehalt in Normal-und Tumorgeweben einiger höherer Pflanzen. *Phytochemistry 15*, 955-56.

Holsters, M.; deWaele, D.; DePicker, A.; Messens, E.; Van Montagu, M.; and Schell, J. (1978) Transfection and transformation of *Agrobacterium tumefaciens. Mol. Gen. Genet. 163*, 181-87.

Kemp, J. D. (1976) Octopine as a marker for the induction of tumorous growth by *Agrobacterium tumefaciens* strain B6. *Biochem. Biophys. Res. Commun. 69*, 816-22.

Kemp, J. D. (1977) A new amino acid derivative present in crown gall tumor tissue. *Biochem. Biophys. Res. Commun. 74*, 862-68.

Kemp, J. D.; Hack, E.; Sutton, D. W.; and El Wakio, M. (1978) Unusual amino acids and their relationship to tumorigenesis. In *Proceedings of the IV International Conference on Plant Pathogenic Bacteria*, Angers, France, pp. 183-88.

Kerr, A. (1969) Transfer of virulence between isolates of *Agrobacterium. Nature 223*, 1175-76.

Kerr, A. (1971) Acquisition of virulence by nonpathogenic isolates of *Agrobacterium radiobacter. Physiol. Plant Path. 1*, 241-46.

Kerr, A.; Manigault, P.; and Tempé, J. (1977) Transfer of virulence *in vivo* and *in vitro* in Agrobacterium. *Nature 265*, 560-61.

Kapwijk, P. M.; Hooykaas, P. J. J.; Kester, H. C. M.; Schilperoort, R. A.; and Rörsch, A. (1976) Isolation and characterization of *Agrobacterium tumefaciens* mutants affected in the utilization of octopine, octopinic acid and lysopine. *J. Gen. Microbiol. 96*, 155-63.

Lederboer, A. (1978) Large plasmids of Rhizobiaceae. Ph.D. thesis, University of Leiden.

Lioret, C. (1957) Les acides aminés libres des tissus de crown-gall cultivés *in vitro*. Mise en évidence d'un acide aminé particulier á ces tissus. *C. R. Acad. Sci. Paris 244*, 2171-74.

Lippincott, B. B.; Whatley, M. H.; and Lippincott, J. A. (1977) Tumor induction of *Agrobacterium* involves attachment of the bacterium to a site on the host plant cell wall. *Plant Physiol. 59*, 388-90.

Ménagé, A., and Morel, G. (1964) Sur la présence d'octopine dans les tissus de crown-gall. *C.R. Acad. Sci. Paris 259*, 4795-96.

Ménagé, A., and Morel, G. (1965) Sur la présence d'un acide aminé nouveau dans le tissu de crown-gall. *C. R. Soc. Biol. Paris 159*, 561-62.

Montoya, A. L.; Chilton, M. -D.; Gordon, M. P.; Sciaky, D.; and Nester, E. W. (1977) Octopine and nopaline metabolism in *Agrobacterium tumefaciens* and crown gall tumor cells: Role of plasmid genes. *J. Bacteriol. 129*, 101-7.

Nevers, P., and Saedler, H. (1977) Transposable genetic elements as agents of gene instability and chromosomal rearrangements. *Nature 268*, 109-15.

Otten, L. A. B. M.; Vreugtenhil, D.; and Schilperoort, R. A. (1977) Properties of D-(+)-lysopine dehydrogenase from crown gall tumor tissue. *Biochem. Biophys. Acta 485*, 268-77.

Palmiter, R. D. (1974) Magnesium precipitation of ribonucleoprotein complexes. Expedient techniques for the isolation of undegraded polysomes and messenger ribonucleic acid. *Biochemistry 13*, 3606-15.

Petit, A., and Tempé, J. (1978) Isolation of *Agrobacterium* Ti-plasmid regulatory mutants. *Mol. Gen. Genet. 167*, 147-55.

Petit, A.; Delhaye, S.; Tempé, J.; and Morel, G. (1970) Recherches sur les guanidines des tissus de crown-gall. Mise en évidence d'une relation biochimique specifique entre les souches d'*Agrobacterium tumefaciens* et les tumeurs qu'elles induisent. *Physiol. Vég. 8*, 205-13.

Petit, A.; Tempé, J.; Kerr, A.; Holsters, M.; Van Montagu, M.; and Schell, J. (1978) Substrate induction of conjugative activity of *Agrobacterium tumefaciens* Ti plasmids. *Nature 271*, 570-72.

Quétier, F., and Vedel, F. (1977) Heterogeneous population of mitochondrial DNA molecules in higher plants. *Nature 268*, 365-68.

Sacristán, M. D., and Melchers, G. (1977) Regeneration of plants from "habituated" and "Agrobacterium transformed" single-cell clones of tobacco. *Mol. Gen. Genet. 152*, 111-17.

Schell, J. (1978) Crown-gall: Transfer of bacterial DNA to plants via the Ti plasmid. In *Nucleic Acids in Plants.* (T. Hall and J. Davies, eds.) (in press).

Sciaky, D.; Montoya, A. L.; and Chilton, M. -D. (1978) Fingerprints of *Agrobacterium* Ti plasmids. *Plasmid 1*, 238-53.

Southern, E. M. (1975) Detection of specific sequences among DNA fragments separated by gel electrophoresis. *J. Mol. Biol. 98*, 503-17.

Turgeon, R.; Wood, H. N.; and Braun, A. C. (1976) Studies on the recovery of crown gall tumor cells. *Proc. Nat. Acad. Sci., U.S.A. 73*, 3562-64.

Van Larebeke, N.; Engler, G.; Holsters, M.; Van den Elsacker, S.; Zaenen, I.; Schilperoort, R. A.; and Schell, J. (1974) Large plasmid in *Agrobacterium tumefaciens* essential for crown gall inducing ability. *Nature 252*, 169-70.

Van Larebeke, N.; Genetello, Ch.; Schell, J., Schilperoort, R. A.; Hermans, A. K.; Hernalsteens, J. P.; and Van Montagu, M. (1975) Acquisition of tumor-inducing ability by non-oncogenic agrobacteria as a result of plasmid transfer. *Nature 255*, 742-43.

Van Larebeke, N.; Genetello, Ch.; Hernalsteens, J.P.; DePicker, A.; Zaenen, I.; Messens, E.; Van Montagu, M.; and Schell, J. (1977) Transfer of Ti plasmids between Agrobacterium strains by mobilization with conjugative plasmid RP4. *Mol. Gen. Genet. 152*, 119-24.

Vedel, F.; Quétier, F.; and Bayen, M. (1976) Specific cleavage of chloroplast DNA from higher plants by Eco RI restriction endonuclease. *Nature 263*, 440-42.

Watson, B.; Currier, T. C.; Gordon, M. P.; Chilton, M. -D.; and Nester, E. W. (1975) Plasmid required for virulence of *Agrobacterium tumefaciens*. *J. Bacteriol.* *123*, 255-64.

White, P. R., and Braun, A. C. (1942) A cancerous neoplasm of plants. Autonomous bacteria-free crown-gall tissue. *Cancer Res.* *2*, 597-617.

Zaenen, I.; Van Larebeke, N.; Teuchy, H.; Van Montagu, M.; and Schell, J. (1974) Supercoiled circular DNA in crown gall inducing Agrobacterium strains. *J. Mol. Biol.* *86*, 109-27.

Part IV: Organelle Transfer

The Analysis of Organelle Behavior and Genetics Following Protoplast Fusion or Organelle Incorporation

H. Bonnett, A. Wallin, and K. Glimelius

Genetic analysis of organellar heredity in plants has been dominated by research on lower plants, such as *Chlamydomonas* and *Saccharomyces*, in which organelles are donated to the zygote from both male and female gametes. Some higher plant species also transfer organelles through both sexes, but, in most plants of agricultural interest, they are transmitted only through the maternal line. Therefore, artificial methods must be devised in which organelles of different genetic composition can be combined within a single cell. In such cells, recombination and segregation of genetic information can be investigated. Several methods involving somatic cells have been developed to yield cells with hybrid organelle populations. These include the following: (1) isolation of chloroplasts from one cell type followed by their introduction into protoplasts of another type, (2) fusion of protoplasts with enucleated protoplasts or protoplasts with impaired nuclei, and (3) fusion of protoplasts.

Note: This research was supported by grant Number BMS 75-13676 from the National Science Foundation and by grants from the American Scandinavian Foundation.

Organelles are not fully autonomous, self-replicating structures but instead depend on the information of the nuclear DNA for many organelle proteins. Moreover, the size of the organellar genome is insufficient to account for all the gene products essential for organelle function. Thus, in organellar hybrids obtained by either of the first two methods, the nucleus must be capable of fulfilling the organellar requirements previously provided by the nucleus of the cell from which the organelle was derived.

The dependence of the organelle on nuclear gene products could conceivably present a hurdle to organelle replication in a foreign nuclear environment. However, recent results analyzing organelle-specific proteins support the contention that nuclear gene products can function in interspecific combinations between organelle and nucleus. The enzyme ribulose bisphosphate carboxylase (RuP_2-case) is the major protein within the chloroplast. It is composed of small subunits coded by nuclear DNA and large subunits coded by chloroplast DNA. This protein will be discussed as an example of synthesis of functional protein molecules from interspecific combinations of nuclear and chloroplast DNA. Then organelle gene products that can be useful in the evaluation of hybrids, including RuP_2-case, will be described, followed by a discussion of recent results applying somatic methods to the field of organelle genetics and segregation.

Interspecific Interactions of Genetic Information in RuP_2-Case Synthesis

Chen and his associates (1976) have analyzed RuP_2-case in more than 60 species of the genus *Nicotiana*. Using isoelectric focusing in polyacrylamide gels, they have separated the large subunits from the small subunits and compared them in each species (Fig. 1). There are 11 different small subunits and 4 different large subunits in various species within the genus. Moreover, in interspecific crosses, small subunits from one species can be combined with large subunits from another species; the enzyme molecules function with a hybrid subunit composition.

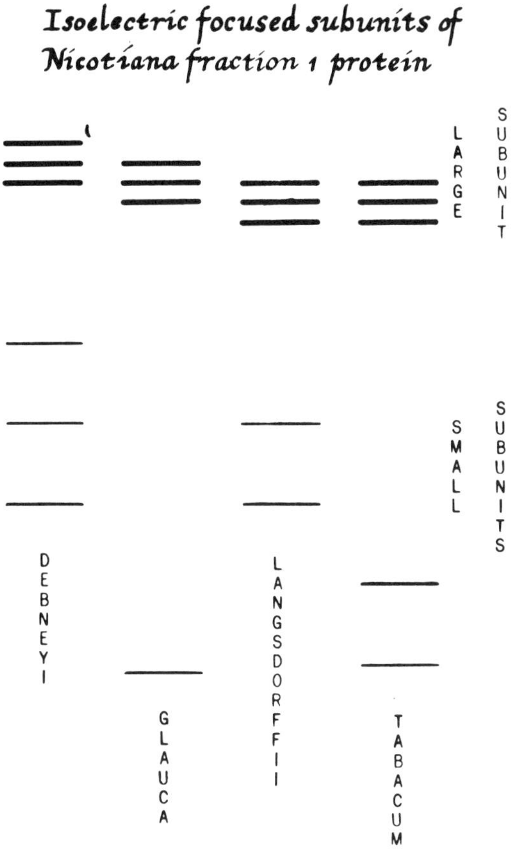

Fig. 1. Diagrammatic representation of the polypeptide composition of RuP_2-case (fraction I protein) from four *Nicotiana* species, following electrofocusing of the isolated protein. The heterogeneity that exists in both the large and small subunits can be utilized as a genetic marker in the analysis of somatic hybrids.

Chua and Schmidt (1978) have provided additional evidence for the conservation of nuclear genetic information necessary for chloroplast function. They used the wheat germ translation system to synthesize the small subunit of RuP_2-case of *Chlamydomonas, Pisum,* and *Spinacia.* In each case, it was synthesized in a precursor form, that is, a polypeptide 4 to 5 kilodaltons larger than the small subunit polypeptide present in the active enzyme. The polypeptide is cleaved to its final size after synthesis on cytoplasmic ribosomes and before association with the larger subunit in the chloroplast. The cleavage and incorporation of the small subunit polypeptide could be carried out by chloroplasts isolated from the same species. In fact, isolated pea chloroplasts could incorporate spinach as well as pea but not *Chlamydomonas* small subunit precursor. The small subunit of spinach could then combine with pea large subunits to make the holoenzyme. Thus, gene products required for the synthesis, processing, and transport of the small subunit precursor could combine from plants as distantly related as pea and spinach.

Such examples provide encouragement that the semiautonomous nature of the chloroplast may be accompanied by sufficient conservation of nuclear information to make chloroplast transplantation between plants of diverse taxonomic groups possible. If transplantation is successful, then the taxonomic limits of organellar-nuclear interdependence can be probed in a way complementary to biochemical analysis of specific proteins.

Organelle-Specific Genetic Characters

Detection and analysis of cells with hybrid organelle populations depend upon the availability of characters that display the genetic information present in the organelles. If the aim of the investigation is to substitute one population of organelles for a genetically different population, then selection systems that act early in cell division and colony formation should be utilized. If, on the other hand, the behavior and segregation of a mixed population of organelles with different genetic backgrounds are to be evaluated, then the procedures used to detect hybrid cells should be based on nuclear characters.

Several characters could be used to select for specific organelles. These include susceptibility to fungal toxins, resistance to antibiotics, and photoautotrophy. *Helminthosporium maydis* toxin quickly kills protoplasts isolated from sensitive strains of corn (Earle et al., 1978). Tentoxin, produced by the fungus *Alternaria alternata,* binds to one subunit of chloroplast-coupling factor that is coded by the chloroplast DNA (Steele et al., 1976). Plants sensitive to tentoxin do not develop a normal photosynthetic apparatus and form white leaves in the presence of the toxin. The effect of the toxin is reversible, since normal leaves will be produced after the toxin is removed.

Maliga (1975) and Dix (1977) and their associates have obtained tobacco cell lines resistant to streptomycin and kanamycin. The genetic basis for this resistance probably resides in the cells' plastids. While antibiotic-resistant cell lines have not yet been utilized in organelle-hybridization studies, they offer the distinct advantage of selection at early stages following hybrid production.

Cells carrying chloroplast defects rendering them to be heterotrophic have been utilized in experiments involving chloroplast transfer by chloroplast uptake (Carlson, 1973; Uchimiya and Wildman, 1979; Landgren and Bonnett, 1979) and by protoplast fusion (Gleba et al., 1975). However, since protoplast cultures must be maintained under heterotrophic conditions until plantlet formation occurs, selection is not possible until many cell generations after hybrid formation.

Additional organelle characters are available that can help distinguish organelle types in cells or plants regenerated from protoplasts, even though they cannot be used as selection agents. These include RuP_2-case, cytoplasmic male sterility (CMS), and organellar DNA. Although significant amounts of RuP_2-case are not produced until the plastids differentiate into chloroplasts, the ability to utilize this marker enzyme as an expression of both nuclear and cytoplasmic DNA in analysis of hybrids accounts for its growing importance.

CMS cannot be detected until the androecium develops. Because CMS plays such an important role in applied genetics and plant breeding, both somatic-hybridization and organelle-transplantation experiments have been undertaken to identify further the cytoplasmic basis for the sterility phenotype.

Finally, unambiguous characterization of the organellar constitution of hybrid tissue or plants can be assessed by electrophoresis of organellar DNA that has been subjected to restriction enzyme degradation. This technique has been shown to be sensitive enough to detect differences at the species level in both chloroplasts (Vedel et al., 1976) and mitochondria (Levings and Pring, 1976), but it requires a substantial amount of tissue to provide sufficient DNA for analysis. The amount of DNA required can be reduced by introducing ^{32}P-labeled phosphate to the tissue to be analyzed. The radioactive organellar DNA can then be restricted and hybridized to DNA isolated from organelles with known DNA-restriction patterns by methods described by Southern (1975).

Organelle Hybrids Produced by Organelle Transplantation

Chloroplasts have been chosen for studies on organelle transplantation since their size and pigmentation permit immediate assessment of incorporation. Mitochondria, by contrast, are small and colorless; there have been no reports of mitochondrial transplantation into plant protoplasts.

Procedures for isolation and *in vitro* maintenance of functional organelles were essential prerequisites to their transplantation into plant protoplasts. Respiratory activity in mitochondria and photosynthetic activity in chloroplasts have been demonstrated through the use of favorable isolation media and gentle techniques of organelle preparation. However, attempts to culture these organelles *in vitro* for long periods or to obtain division of organelles have failed. The most stable organelles in culture have been chloroplasts isolated from siphonaceous algae (Giles and Sarafis, 1972). Chloroplasts from such algae are also borrowed by Sacoglossan molluscs. These animals incorporate chloroplasts of both green (Trench, 1969) and red (Kremer and Schmitz, 1976) algae into their digestive cells, where the chloroplasts retain their photosynthetic integrity for at least several weeks and contribute significantly to the fixed carbon requirements of the animal. This chloroplast symbiosis demonstrates that chloroplasts can be removed from the plant and introduced into a foreign cytoplasm without loss of function.

Chloroplasts from several plant species have been incorporated into protoplasts using polyethylene glycol (Bonnett and Eriksson, 1974; Vasil and Giles, 1975; Davey et al., 1976). Although the conditions that favor incorporation are similar, if not identical, to those that promote protoplast fusion, the mode of entry of the chloroplast is uncertain. Davey and his colleagues (1976), using protoplasts isolated from *Parthenocissus* and chloroplasts from *Petunia*, concluded that chloroplast uptake is a fusion process involving the chloroplast outer membranes and the protoplast membrane. In such a case, only the interior thylakoid and the stromal components of the chloroplasts are introduced into the protoplast. Bonnett (1976) concluded that at least some chloroplasts isolated from the alga *Vaucheria* gained entry to carrot protoplasts with intact outer membranes.

There are still very few reports concerning the ability of the incorporated chloroplasts to function and replicate or the ability of the protoplast to survive following the transplantation. Using *Daucus* protoplasts isolated from cells in suspension culture and chloroplasts isolated from the alga *Vaucheria*, the growth of protoplasts containing chloroplasts has been followed in drop cultures. Figs. 2a, 2b, and 2c are a series of photomicrographs of a *Daucus* protoplast containing algal chloroplasts. The protoplast has formed a cell wall and the chloroplasts were observed to move around in the cell over the course of a 4-day period. Although this protoplast did not divide, other protoplasts containing chloroplasts have undergone cell division (Figs. 2d and 2e). Similar observations have been made on albino tobacco protoplasts containing chloroplasts transplanted from *Nicotiana glauca*. Observations on such cells showed that wall regeneration and chloroplast movement combine to make impossible an accurate assessment of the number of chloroplasts in a cell. Fluorescence microscopy frequently revealed chloroplasts that could not be seen by brightfield microscopy due to obscuring cytoplasmic structures. However, chloroplasts bleached quickly following fluorescence-microscope examination. Because of these technical problems, a quantitative analysis of chloroplast movement or replication during protoplast-cell-wall regeneration and division has not been possible.

The goal of producing a plant in which specific characters of transplanted chloroplasts are expressed has eluded investigators except for the report by Kung and his associates (1975) that chloroplasts isolated from *N. suaveolens* and introduced into albino

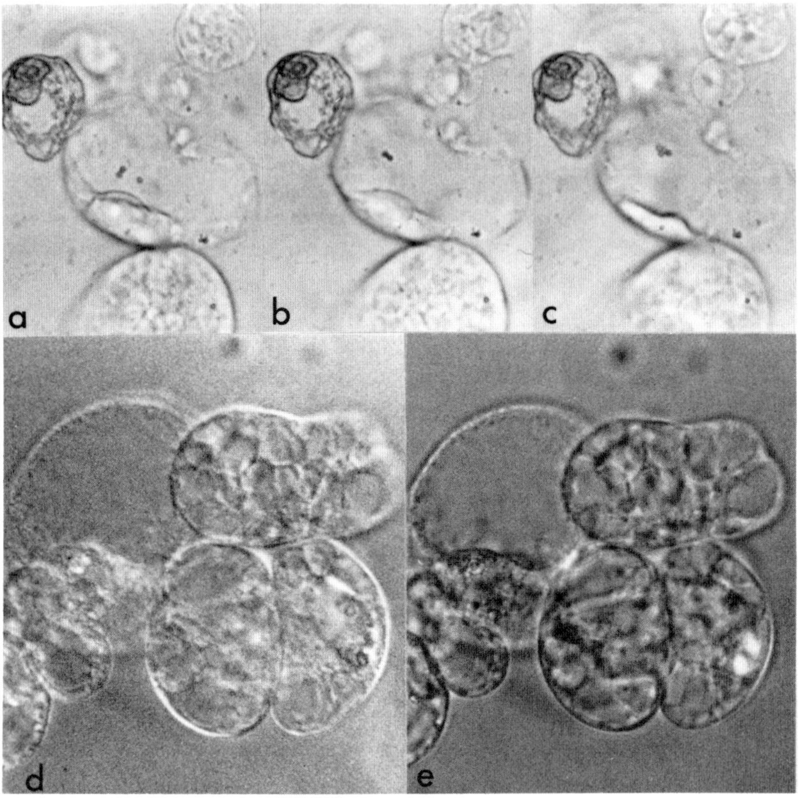

Fig. 2. Photomicrographs of *Daucas* protoplasts containing chloroplasts isolated from the alga *Vaucheria*. (a) (×500) Protoplast cultured for 3 days containing five chloroplasts, of which three are shown together in the parietal cytoplasm. (b) Same protoplast a few hours later, showing one of the three chloroplasts has moved away. (c) Same protoplast after 4 days in culture; now four chloroplasts are aggregated in the same region of the parietal cytoplasm. (d) Two cells (×1000) that have divided from a single protoplast after 2 days in culture. Two chloroplasts are visible in one of the daughter cells. (Nomarski optics.) (e) Identical view of the cells taken with fluorescence epi-illumination and phase-transmitted illumination. The two chloroplasts are shown as bright, fluorescent organelles.

N. tabacum protoplasts led to regeneration of a green plant containing the large subunit of RuP_2-case characteristic of *N. suaveolens.* This plant also had the small subunit of both *N. tabacum* and *N. suaveolens,* indicating that at least some of the *N. suaveolens* nuclear genome was in the plant as well. The authors suggested that both a nucleus and a chloroplast could have been incorporated. However, the plant could have been instead a chimera regenerated from protoplasts of each *Nicotiana* species.

More recently, Uchimiya and Wildman (1979) have tried to repeat this experiment, using chloroplasts from *N. gossei* and protoplasts derived from albino tobacco pith tissue. RuP_2-case with large subunit from *N. gossei* could be distinguished by immunological electrophoresis from that of *N. tabacum.* Both green calli and green shoots, obtained from protoplasts that had been treated with chloroplasts in the presence of polyethylene glycol, were analyzed to search for the expression of the *N. gossei* chloroplasts. None was found. Landgren and Bonnett (1979) treated mixtures of albino tobacco-mesophyll protoplasts and chloroplasts isolated from *N. excelsior* with polyethylene glycol. All the plants regenerated from these protoplasts were heterotrophic, indicating a failure of transplanted *N. excelsior* chloroplasts to lead to green-plant formation.

The inability to obtain evidence for function and replication of transplanted chloroplasts may result from several causes. Perhaps the uptake process is so disruptive to the chloroplast that it is damaged beyond replication. This difficulty could be overcome by improvements in existing methods or by protecting the chloroplasts by enclosing them in artificial lipid membranes (Giles, 1978). Alternatively, the progeny of one or two chloroplasts, introduced into a protoplast full of genetically defective albino plastids, may not segregate in a way fortuitous enough to lead to green plantlets. In this case, employing a positive selection technique, such as utilizing chloroplasts carrying a resistance marker for antibiotics (Dix et al., 1977), could lead to success. In addition, the number of chloroplasts introduced into a single protoplast could be increased. Giles (1978) has used liposomes to introduce more chloroplasts into a single protoplast than results from the use of polyethylene glycol. No information is available on either the survival of the protoplasts or their regeneration following fusion with liposome vesicles containing chloroplasts.

Organelle Hybrids Produced by Cell Fusion

Organelles can be introduced by protoplast fusion into a protoplast containing a nucleus foreign to them. In a recent technique developed by Wallin and her associates (1978), cytochalasin B and high centrifugal forces were utilized to separate protoplasts into miniprotoplasts, which contain the nucleus and some organelles, and subprotoplasts, which contain organelles but no nuclei. So far, these subprotoplasts have not been employed in fusion experiments designed to yield cytoplasmic hybrids.

Cytoplasmic hybrids have been produced by use of X-irradiated protoplasts (Zelcer et al., 1978). At X-ray dosages that prevent replication of the nuclear genetic information but that do not prevent division of mitochondria and chloroplasts, organelles can be transferred to an unirradiated protoplast by protoplast fusion. Protoplasts of a male-sterile cultivar of *N. tabacum* containing organelles derived from *N. suaveolens* were irradiated and fused with unirradiated protoplasts from *N. silvestris*. Plants were recovered that expressed *N. silvestris* morphology but also displayed the male-sterility trait.

Compatibility and segregation of chloroplast types have been assessed in a few cases through analyses of hybrids produced by somatic hybridization. Smith, Kao, and Combatti (1976) reared 16 different hybrid plants resulting from fusions induced between protoplasts of *N. glauca* and *N. langsdorffii*. The chloroplast complement of these plants was assayed by isoelectric focusing of RuP_2-case isolated from the hybrids (Chen et al., 1977). They reported that 8 hybrids had *N. glauca* type large subunits, 7 had *N. langsdorffii* type, and 1 unhealthy plant had both. The unhealthy plant was subsequently shown to be a chimera rather than a hybrid.

The hybrid plants Melchers and his associates (1978) recently obtained between tomato and potato also showed one or the other chloroplast large subunit type, but not both. In the limited number of cell divisions necessary to progress from a hybrid protoplast to the organization of plantlets, the chloroplast populations were segregating exclusively but apparently without bias; no clear competitive advantage was displayed by either parental type chloroplast. Whereas total chloroplast segregation by chance alone is very unlikely for the hybrid plants produced in these experiments, additional experiments will be necessary before judging the universality of this pattern of organelle segregation in somatic hybrids.

Analysis of Cytoplasmic Male Sterility by Protoplast Fusion

CMS in the S cytoplasm of corn has been correlated with the presence of a DNA molecule found within the mitochondria but separate from the mitochondrial genome (Levings and Pring, 1978). In some cultivars of wheat, CMS has been correlated with the DNA contained within the mitochondria (Quetier and Vedel, 1977). However, in a paper on phenotypic expression of chloroplast DNA in tobacco Chen and his associates (1977a) observed that the pattern of inheritance of RuP_2-case was correlated with CMS in tobacco and therefore suggested that the sterility trait may result from nuclear-chloroplast DNA interaction. Male-sterile cultivars in tobacco are usually derived by interspecific hybridization and are known to exhibit a wide range of sterile phenotypes, ranging from petaloid development of the anthers or nearly complete absence of the androecium to normal development through meiosis followed by abortion of the microspores (Fig. 3).

Attempts to prove that chloroplast DNA is involved in male sterility by direct chloroplast transfer in tobacco have not succeeded (Uchimiya and Wildman, 1979; Landgren and Bonnett, 1979). Beillard and her associates (1978) have analyzed restriction patterns of chloroplast DNA and CMS in plants selected from a somatic-hybridization experiment involving two cultivars of *N. tabacum*, one male fertile and the other male sterile, with *N. debneyi* cytoplasm. Plants were selected from the hybridization experiment that demonstrated the nuclear character of one or the

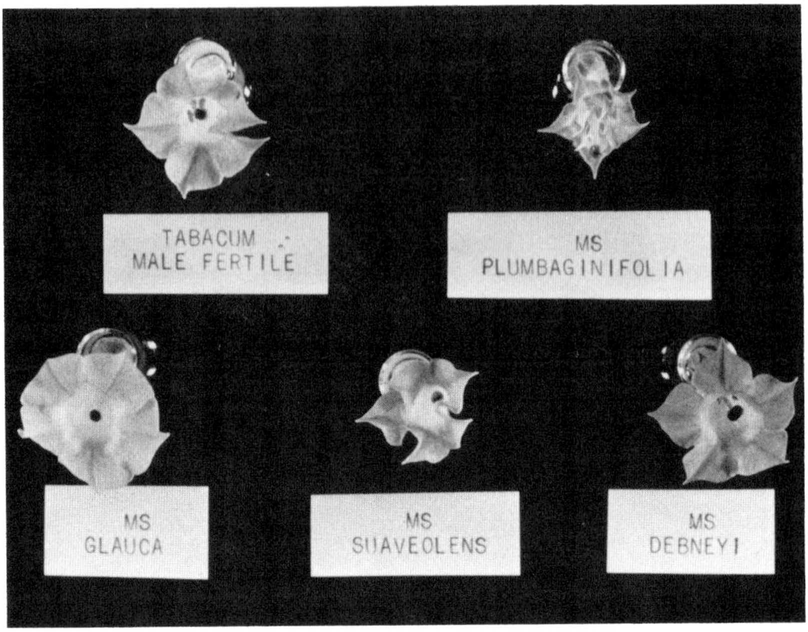

Fig. 3. Male-sterile flowers from cultivars of *N. tabacum* showing variations in stamen development. Stamens show a petaloid character in the male-sterile *plubaginifolia* cultivar. Stamens are absent in the *glauca* and *suaveolens* cultivars. The male-sterile *debneyi* cultivar has one chromosome derived from *N. debneyi* that restores normal stamen development through meiosis, but then the developing pollen grains abort. Natural size.

other parent and that were also intermediate in the male-sterile phenotype. The DNA of the hybrids, isolated from chloroplasts and treated with restriction enzymes, produced electrophoretic patterns characteristic of either the *debneyi* or the *tabacum* chloroplast complement. No evidence was found that these plants contained a mixed population of chloroplasts. Hybrid plants were found to exhibit male sterility that had a *tabacum* chloroplast complement; therefore, male sterility was no longer strictly associated with the *debneyi* chloroplast complement. However, in those plants in which a distinct nuclear character, which had been associated with the fertile parent, became associated with a severe sterile phenotype following somatic hybridization, the chloroplast type was altered from the *tabacum* to the *debneyi* type. Similarly, in those plants in which another distinct nuclear character, which had been associated with the sterile parent, became associated with a fully fertile phenotype, the chloroplast type was altered from the *debneyi* to the *tabacum* type. Thus, in those cases in which the sterility phenotype changed most dramatically with respect to the nuclear genome, there was also a change in the chloroplast type as a result of the fusion process.

Male sterility in a tobacco cultivar derived from interspecific hybridization of *N. tabacum* and *N. debneyi* has been studied by Sand and Christoff (1973). They described male-sterile plants ranging from those with highly split corollas and stigmas in place of anthers to plants with normal sympetalous corollas and nearly normal (but sterile) anthers. This variation in male-sterile-phenotypic expression in the *N. debneyi* cultivar confuses its use for the detection of cytoplasmic hybrids resulting from somatic hybridization experiments.

Some plants obtained by Beillard and her associates (1978) showed more than one flower type on the same plant, although these plants were not used in the DNA analyses. They point out that the analysis of mitochondrial DNA in such experiments will help clarify the role of organellar DNA in CMS. The segregation of mitochondria from hybrid cytoplasm as well as the dependent or independent assortment of mitochondrial and chloroplast populations in hybrid cells remains to be investigated.

Gleba (1978) hybridized albino protoplasts from a male-fertile cultivar of tobacco with a male-sterile cultivar also deriving its cytoplasmic organelles from *N. debneyi*. He selected variegated plantlets and found that they subsequently exhibited the male-sterile phenotype. These experiments represent the only report so far in which two genetically different plastids (albino mutant plastids from *tabacum* and autotrophic *debneyi* chloroplasts) coexist in plants regenerated from somatic-hybridization experiments. In other experiments, selection for hybrids on the basis of nuclear characters may have discriminated against somatic plants with hybrid cytoplasms. Since such variegations occasionally sort out to produce green shoots and white shoots, the expression of fertility in totally green or white shoots would be of interest with respect to the involvement of the chloroplast genome. A better understanding of those combinations of nuclear and organellar genomes that lead to failure of normal male-gamete formation should provide a focus for deriving male-sterile cultivars using protoplast techniques.

Conclusions

Somatically produced organelle hybrids offer great possibilities for the analysis of organelle behavior and genetics. Since most plants show uniparental patterns of organelle transmission from generation to generation, somatic techniques have to be used to yield organelle heterozygotes in which assortment and recombination of genetic information between nuclear and organellar genetic material can be investigated.

The genetic information present in organellar DNA controls biosynthetic steps of fundamental importance to the plant, such as pollen formation and carbon dioxide fixation. If protoplast techniques can be further improved to facilitate organelle transfer consistently, then the variations in function of those gene products partially coded by organelles can be effectively used in plant breeding.

References

Beillard, G.; Pelletier, G.; Vedel, F.; and Quetier, F. (1978) Morphological characteristics and chloroplast DNA distribution in different cytoplasmic parasexual hybrids of *Nicotiana tabacum. Mol. Gen. Genet. 165,*231-37.

Bonnett, H. T. (1976) On the mechanism of the uptake of *Vaucheria* chloroplasts by carrot protoplasts treated with polyethylene glycol. *Planta 131*, 229-33.

Bonnett, H. T., and Eriksson, T. (1974) Transfer of algal chloroplasts into protoplasts of higher plants. *Planta 120*, 71-79.

Carlson, P. S. (1973) The use of protoplasts for genetic research. *Proc. Nat. Acad. Sci., U.S.A., 70*, 598-602.

Chen, K.; Johal, S.; and Wildman, S. G. (1976) Role of chloroplast and nuclear DNA genes during evolution of fraction I protein. In *Genetics and Biogenesis of Chloroplasts and Mitochondria* (T. Büchner, W. Neuport, W. Sebald, and S. Werner, eds.), pp. 3-11. North Holland Publishing Company, Amsterdam.

Chen, K.; Johal, S.; and Wildman, S. G. (1977a) Phenotypic markers for chloroplast DNA genes in higher plants and their use in biochemical genetics. In *Nucleic Acid and Protein Synthesis in Plants* (J. W. Weil and L. Bogorad, eds.), pp. 183-94. Plenum Publishing Company, New York.

Chen, K.; Wildman, S. G.; and Smith, H. H. (1977b) Chloroplast DNA distribution in parasexual hybrids as shown by polypeptide composition of fraction I protein. *Proc. Nat. Acad. Sci., U.S.A., 74*, 5109-12.

Chua, N. H., and Schmidt, G. W. (1978) Post-translational transport into intact chloroplasts of a precursor to the small subunit of ribulose-1,5-biphosphate carboxylase. *Proc. Nat. Acad. Sci., U.S.A., 75*, 6110-14.

Davey, M. R.; Frearsson, E. M.; and Power, J. B. (1976) Polyethylene glycol-induced transplantation of chloroplasts into protoplasts: An ultrastructural assessment. *Plant. Sci. Lett. 7*, 7-16.

Dix, P. J.; Joo, F.; and Maliga, P. (1977) A cell line of *Nicotiana sylvestris* with resistance to kanamycin and streptomycin. *Mol. Gen. Genet. 157*, 285-90.

Earle, E. D.; Gracen, V. E.; Yoder, O. C.; and Gemmill, U. P. (1978) Cytoplasm-specific effects of *Helminthosporium maydis* race T toxin on survival of corn mesophyll protoplasts. *Plant Physiol. 61*, 420-24.

Giles, K. L. (1978) The uptake of organelles and microorganisms by plant protoplasts: Old ideas but new horizons. In *Frontiers of Plant Tissue Culture/1978* (T. A. Thorpe, ed.), pp. 67-74. International Association for Plant Tissue Culture, Calgary, Alberta, Canada.

Giles, K. L., and Sarafis, V. (1972) Chloroplast survival and division *in vitro*. *Nat. New Biol. 236*, 56-58.

Gleba, Y. Y. (1978) Extranuclear inheritance investigated by somatic hybridization. In *Frontiers of Plant Tissue Culture 1978* (T. A. Thorpe, ed.), pp. 95-102. International Association for Plant Tissue Culture, Calgary, Alberta, Canada.

Gleba, Y. Y.; Butenko, R. G.; and Sytnik, U. M. (1975) Fusion of protoplasts and parasexual hybridization in *Nicotiana tabacum* L. *Dokl. Akad. Nauk. U.S.S.R. 221*, 1196-98.

Kremer, B. P., and Schmitz, K. (1976) Rhodoplasts functional in snails. *Marine Biol. 34*, 313-16.

Kung, S. D.; Gray, J. C.; Wildman, S. G.; and Carlson, P. S. (1975) Polypeptide composition of fraction 1 protein from parasexual hybrid plants in the genus *Nicotiana*. *Science 187*, 353-55.

Landgren, C. R., and Bonnett, H. T. (1979) The culture of albino tobacco protoplasts treated with polyethylene glycol to induce chloroplast incorporation. *Plant Sci. Lett. 16*, 15-22.

Levings, C. S., III, and Pring, D. R. (1976) Restriction endonuclease analysis of mitochondrial DNA from normal and Texas cytoplasmic male-sterile maize. *Science 193*, 158-60.

Levings, C. S., III, and Pring, D. R. (1978) The mitochondrial genome of higher plants. *Stadler Symp.* (G. P. Redei, ed.), pp. 77-93.

Maliga, P.; Sz-Breznovits, A.; and Marton, L. (1975) Non-Mendelian streptomycin-resistant tobacco mutant with altered chloroplasts and mitochondria. *Nature 255*, 401-2.

Melchers, G.; Sacristan, M. D.; and Holder, A. A. (1978) Somatic hybrid plants of potato and tomato regenerated from fused protoplasts. *Carlsberg Res. Commun. 43*, 203-18.

Quetier, F., and Vedel, F. (1977) Heterogenous population of mitochondrial DNA molecules in higher plants. *Nature 268*, 365-68.

Sand, S. A., and Christoff, G. T. (1973) Cytoplasmic-chromosomal interactions and altered differentiation in tobacco. *J. Heredity 64*, 24-30.

Smith, H. H.; Kao, K. N.; and Combatti, N. C. (1976) Confirmation and extension of interspecific hybridization by somatic protoplast fusion in *Nicotiana. J. Heredity 67*, 123-28.

Southern, E. M. (1975) Detection of specific sequences among DNA fragments separated by gel electrophoresis. *J. Mol. Biol. 98*, 503-17.

Steele, J. A.; Uchytil, T. F.; Durbin, R. D.; Bhatnagar, P.; and Rich, D. H. (1976) Chloroplast coupling factor 1: A species-specific receptor for tentoxin. *Proc. Nat. Acad. Sci., U.S.A., 73*, 2245-48.

Trench, R. K. (1969) Chloroplasts as functional endosymbionts in the mollusc, *Trichadia crispata* (bergh. Saccoglossa). *Nature 222*, 1071-72.

Uchimiya, H., and Wildman, S. G. (1979) Non-translation of foreign genetic information for fraction 1 protein under circumstances favorable for direct transfer of *Nicotiana gossei* isolated chloroplasts into *N. tabacum* protoplasts. *In Vitro 15*, 463-68.

Vasil, I. K., and Giles, K. L. (1975) Induced transfer of higher plant chloroplasts into fungal protoplasts. *Science 190*, 680.

Vedel, F.; Quetier, F.; and Bayen, M. (1976) Specific cleavage of chloroplast DNA from higher plants by Eco RI restriction nuclease. *Nature 263*, 440-42.

Wallin, A.; Glimelius, K.; and Eriksson, T. (1978) Enucleation of plant protoplasts by cytochalasin B. *Z. Pflanzenphysiol. 87*, 333-40.

Zelcher, A.; Aviv, D.; and Galun, E. (1978) Interspecific transfer of cytoplasmic male sterility by fusion between protoplasts of normal *Nicotiana sylvestris* and X-ray irradiated protoplasts of male sterile *N. tabacum. Z. Pflanzenphysiol. 90*, 397-407.

Chromosome-Mediated Gene Transfer and Microcell Hybridization

R. S. Athwal and O. W. McBride

Somatic cell genetics has provided a highly productive approach to the genetic analysis of mammalian cells. An interesting feature of this method is its ability to combine genotypes by cell hybridization beyond species boundaries. Thus, intraspecies variation, essential for conventional genetic analysis in eukaryotic organisms, is substituted by interspecies variation.

The most useful approach for genetic analysis of eukaryotic cells requires that the quantity of transferred genetic information be selected and varied from a single gene to an entire genome. The array of techniques now available for transfer of genetic material appear to approach this ideal.

Somatic cell hybridization can be considered as the transfer of the total genome from one cell to another. Usually, cells of two different species are fused, and one of the input parental chromosome sets is subsequently partially eliminated from the resultant hybrid cells. The chromosomes segregate largely on a random basis. Thus, a series of hybrids can be generated that possess different subsets of chromosomes derived from one of the parental species. The analysis of these segregated hybrids permits the assigment of genes to specific chromosomes. The important methodological contributions and major results obtained by this method have been thoroughly discussed in numerous reviews (Ruddle and Creagan, 1975; McKusick and Ruddle, 1977).

Recently, two additional techniques have been developed to transfer a limited amount of genetic information between eukaryotic cells, namely chromosome-mediated gene transfer (McBride and Ozer, 1973a) and microcell-mediated chromosome transfer (Ege and Ringertz, 1974). Both of these techniques permit the selective transfer of a limited portion of the donor genome. In this review, we will discuss these two methods in detail. Selectable markers can also be transferred by the cellular uptake of purified deoxyribonucleic acid (DNA) (Wigler et al., 1978), but that is beyond the scope of this paper.

Chromosome-Mediated Gene Transfer

Metaphase chromosomes represent a biological fractionation of the eukaryotic genome and permit the transfer of individual linkage groups of one cell type to another. Since the first demonstration of functional gene transfer by the uptake of isolated metaphase chromosomes (McBride and Ozer, 1973a), this method has been utilized in several laboratories. The results of gene-transfer experiments have been reviewed (McBride and Athwal, 1976, 1977, 1979; Willecke, 1978) and are listed in Table 1. This technique will be summarized and further details can be found in the other reviews.

Chromosome-mediated gene transfer (CMGT) requires incubation of isolated metaphase chromosomes with recipient cells and selection for gene-transfer clones in appropriate media following chromosome uptake. Many methods have been described for the isolation and purification of metaphase chromosomes (Mendelson et al., 1968; Maio and Schildkraut, 1969; Wray et al., 1972; McBride and Ozer, 1973b). Usually, a synchronized population of cells is arrested in mitosis by addition of colchicine or colcemid. The cells are centrifuged, washed, and swollen in a hypotonic buffer such as 0.075 M KCl. The cells are disrupted and metaphase chromosomes are released by homogenization in a blender. The

Table 1. Frequency of Gene Transfer

Donor[a]	Recipient[a]	Selection	Marker	Number of colonies[b]	Frequency ($\times 10^{-7}$)	Reference[d]
1. CH	$M(A_9)$	HAT	HPRT	10	1.0	1
2. CH	CH	HAT	HPRT	6	1.1[c]	2
3. H	$M(A_9)$	HAT	HPRT	4	0.6	3
4. H	$M(A_9)$	HAT	HPRT	3	2.5	4
5. CH	$M(B_{82})$	HAT	TK	12	2.4	5
6. H	$M(B_{82}, LMTK^-)$	HAT	TK	8	1.6	6
7a. CH-SV40	BSC-1	–	T-antigen SV40	–	$(1\text{-}2.5 \times 10^3)$	7
7b. CH-SV40	BSC-1	–	Virus rescue	–	25	7
8. CH-SV40	BSC-1	–	Virus rescue	–	4-42	8
9a. CH	CH	HAT	HPRT	(>1000)	490-1660[c]	9
9b. H	CH	HAT	HPRT	30	30	9
10. H	HxCH[e]	HAT	HPRT	–	135-320	10
11. H	CH	HAT	TK	–	107-660	11
12. M	$M(A_9)$	HAT	HPRT	1	0.3	12
13. M(H-HPRT)	CH	HAT	HPRT	5	0.9	13
14a. $CH(Mtx^{RIII}Ova^R)$	CH	Mtx	Mtx^{RIII}	49	150	14[f]
14b. $CH(Mtx^{RIII}Ova^R)$	CH	Ova	Ova^R	79	160[c]	
15a. $CH(Mtx^{RIII}Gat^+)$	$CH(Mtx^SGat^-)$	*a*	Gat	56	140[c]	15[f]
15b. $CH(Mtx^{RIII}Gat^+)$	$CH(Mtx^SGat^-)$	*a* + Mtx	Gat + Mtx^{RIII}	29	70[c]	
15c. $CH(Mtx^{RIII}Gat^+)$	CH-Gly A^-	*a*	Gly A	13	65[c]	
15d. $CH(Mtx^{RIII}Gat^+)$	CH-Gly B^-	*a*	Gly B	12	60[c]	
15e. $CH(Mtx^{RIII}Gat^+)$	CH-Gly B^-	*a* + Mtx	Gly B + Mtx	11	55[c]	
16a. CH	SH	agar	aga^+	>100	200	16[f]
16b. CH	CH	agar	aga^+	15	50	
17a. H	$M(A_9)$	HAT	HPRT	≥42	30	17[f]
17b. H	$M(A_9)$	HAT	HPRT	≥265	400	

Donor[a]	Recipient[a]	Selection	Marker	Number of colonies[b]	Frequency ($\times 10^{-7}$)	Reference[d]
18a. H	M(3T3-4E)	HAT	TK	–	50	18[f]
18b. H-HSV-1	M(3T3-4E)	HAT	TK	>3	2.5	
18c. M-HSV-1	M(3T3-4E)	HAT	TK	–	2.5	
19. M-HSV-1	M(LMTK⁻)	HAT	TK	4	–	19[f]
20. H	M(LMTK⁻)	HAT	TK	7	2.3	20

[a] Chromosomal donor and recipient species: CH = Chinese hamster; M = mouse; H = human; BSC-1 = monkey; CH-SV40 = transformed Chinese hamster cells; SH = Syrian hamster; M(or H)-HSV-1 = mouse (or human) cells containing integrated Herpes simplex viral type I DNA.

[b] Gene product was characterized as chromosomal donor species in most cases.

[c] Chromosomal donor and recipient species of gene product were indistinguishable.

[d] References: (1) McBride and Ozer, 1973a; (2) McBride and Ozer, 1973b; (3) Burch and McBride, 1975; Willecke and Ruddle, 1975; (5) Ruddle and McBride, 1977; (6) Willecke et al., 1976a; (7) Shani et al., 1974; (8) Shani et al., 1976; (9) Wullems et al., 1975; (10) Wullems et al., 1976b; (11) Wullems et al., 1977; (12) Degnen et al., 1976; (13) Athwal and McBride, 1977; (14) Spandidos and Siminovitch, 1977a; (15) Spandidos and Siminovitch, 1977a; (15) Spandidos and Siminovitch, 1977b; (16) Spandidos and Siminovitch, 1977c; (17) Miller and Ruddle, 1978; (18) Sabourin and Davidson, 1979; (19) Scangos et al., 1979; (20) Klobutcher and Ruddle, 1979.

[e] Reduced human × CH hybrid containing about four human chromosomes.

[f] Chromosomes applied to monolayer of cells as $CaPO_4$ precipitate.

[g] Posttreatment with dimethyl sulfoxide in 17b.

chromosomes are further purified from contaminating nuclei, cell debris, and intact cells by differential, isopycnic, and unit-gravity sedimentation (McBride and Ozer, 1973b).

These partially purified chromosomes are mixed with recipient cells in suspension at high concentration (10^7 cells/ml) in multiplicities of one donor chromosome set per recipient cell and incubated at 37° C for two to three hours, with gentle mixing. The suspension is then inoculated into petri dishes (10^4 cells/cm^2) in minimal essential medium.

Following growth of the cells in nonselective medium for three cell generations to allow time for expression of the transferred phenotype, the growth medium is replaced with a selective medium, and then the plates are refed twice a week for several weeks. Colonies that appear during selection are isolated from separate plates. The progeny of isolated colonies are further analyzed biochemically for expression of the transferred phenotypes. Each of these colonies represents the progeny of an independent gene-transfer event.

The frequency of transfer of any single functional gene by this method is relatively low (10^{-6}-10^{-7}/cell). Thus, the recovery of the progeny of individual gene-transfer events (transgenotes) from a population of recipient cells requires a sensitive selection system. Cell lines carrying enzyme-deficiency mutations are usually used as recipients, and selection for gene transfer is carried out under nonpermissive nutritional conditions. Thus, survival of any cell under these selective conditions is dependent on complementation of the mutation by a transferred donor chromosome (transgenome). The HAT (hypoxanthine, amethopterin, and thymidine) selection system (Szybalska and Szybalski, 1962; Littlefield, 1964) is most frequently used. However, other sensitive selection methods have also been employed (Spandidos and Siminovitch, 1977a, 1977b, 1977c, 1978; Shani et al., 1974, 1976).

Since the frequency of gene transfer is in the same range as spontaneous mutation frequencies, it is essential to distinguish gene-transfer clones from revertants. Several different criteria have been employed for this purpose. The identification of the transferred gene (transgenome) or gene product as chromosomal donor species is the most reliable criterion for authentic gene transfer. Hence, cells with marker phenotypes that differ detectably are used as donors and recipients. This usually requires the use of an interspecific gene-transfer system. Both simple comparison of the relative frequency of colony formation following CMGT with spontaneous mutation rates of the recipient cells and instability of the transferred phenotype in the absence of selection also have been employed as criteria for transferrents in intraspecific combinations (Wullems, 1975; Spandidos and Siminovitch, 1977a).

Chromosome Uptake and Frequency of Gene Transfer

The mechanism of chromosome uptake and functional gene transfer is poorly understood, but the following is a generally accepted model. Donor chromosomes probably enter the recipient cells by endocytosis (McBride and Athwal, 1976; McBride and Ozer, 1973b; McBride, 1974). A majority of internalized chromosomes are degraded by lysosomal enzymes, but larger fragments of a chromosome may escape occasionally from a phagosome and be incorporated into the nucleus at mitosis. Thus, the incorporation and expression of a gene by CMGT can be viewed as a two-step process involving phagocytic uptake of a chromosome and its escape from the phagosome into an environment conducive to its expression. The frequency of functional gene transfer is dependent upon a number of factors affecting these two processes.

Modification of the procedure for incubating chromosomes with recipient cells has been demonstrated to increase dramatically the frequency of functional gene transfer. Miller and Ruddle (1978) applied human chromosomes as a calcium phosphate precipitate to monolayers of mouse A_9 cells and selected for hypoxanthine phosphoribosyl transferase *(hprt)* transfer in HAT medium. They observed about a ten-fold increase in the frequency of functional gene transfer ($2\text{-}3 \times 10^{-6}$) under these conditions. The transfer frequency was further increased by another order of magnitude (4×10^{-5}) when the recipient cells were exposed to 10% dimethyl sulfoxide for one-half hour at an interval of four hours following chromosome uptake. This increase in frequency of transfer also correlated with an increased size of the transferred fragment (transgenome) as determined by the frequency of co-transfer of syntenic markers and detection of microscopically visible chromosome fragments.

Several other factors may also influence the frequency of gene transfer. Wullems et al. (1975) reported a ten times higher frequency in intraspecific gene-transfer combinations. However, this interpretation was not supported by results from another laboratory (Degnen et al., 1976) using an intraspecies combination in which the donor phenotype differed from the recipient. Transferrents could be distinguished from revertants in the latter study.

Transfer frequencies may also be affected by the stage of the cell cycle of recipient cells during uptake. A higher transfer frequency (approximately ten-fold) has been reported (Wullems et al., 1975) when mitotic cells were used as recipients. It is possible that, in mitotic cells, a donor chromosome could be directly incorporated with host chromosomes into the nucleus of the cell at teleophase. Increase in transfer frequency might also result from depression of the activity of degradative enzymes. The reported experiments involved intraspecific transfer, so the species of the transferred gene product could not be identified as donor. In contrast to these results, mitotic cells have been reported to bind, or ingest, very few chromosomes (McBride, 1974). Also, the specific activities of lysosomal enzymes such as acid DNase, glucoronidase, N-acetyl glucoseaminidase, galactosidase, and cathepsin D have been found to remain constant throughout the cell cycle (Berg et al., 1975).

In recent experiments involving the transformation of cells with purified DNA (Wigler et al., 1978), it has been found that the frequency of transfer of a specific marker varies with different recipient cell lines (Peterson and McBride, 1979). This finding suggests that similar differences in cell lines may exist for the uptake and expression of genes by CMGT. No systematic study has been made to determine the effect of different recipient cell lines. This could be accomplished using single-donor chromosome preparations for the transfer of a particular marker in combination with a number of recipient cell lines.

Size of the Transgenome

The size of the transgenome has been assessed by indirect means and found to be smaller than an intact chromosome. Following transfer of the human hprt gene to mouse A_9 cells (Burch and McBride, 1975; Willecke and Ruddle, 1975), colonies of gene-transfer cells (transgenotes) were isolated in HAT medium. Whereas the human form of the selected HPRT marker was demonstrated in the extracts of these colonies, the human form of the flanking markers, glucose-6-phosphate dehydrogenase (G6PD) and phosphoglycerate kinase (PGK), was not detected, nor were two other X-chromosomal markers galactosidase and a human species-specific X-linked surface antigen, SAX (McBride and Athwal, 1976; Willecke, 1978). A donor chromosome or fragment was also not detectable in these gene-transfer clones cytologically. Based upon the known size of the human X chromosome and the regional X-chromosomal assignments, the distance between G6PD and PGK is about 1% of the haploid human genome. Thus, it was estimated that the transgenome must be smaller than 1% of the human genome (Burch and McBride, 1975; Willecke and Ruddle, 1975; McBride and Athwal, 1976; Willecke, 1978).

A better estimate of transgenome size was obtained by transfer of the thymidine kinase gene *(tk)* (McBride and Athwal, 1976; Willecke et al., 1976c; Ruddle and McBride, 1977; McBride et al., 1978). This gene is closely linked to the galactokinase locus (Galk) in humans as well as in other species (Elsevier et al., 1974; Kozak and Ruddle, 1977; McBreen et al., 1977; Ruddle and McBride, 1977). Cotransfer of galactokinase by CMGT was observed when selection was carried out only for the transfer of *tk*. The frequency of cotransfer of these two linked loci *(tk* and *galk)* has been used to estimate the size of the transgenome. Since the cotransfer of these two loci occurs in 20% to 25% of the clones isolated for the transfer of *tk*, the transferred fragment has been estimated to be 25% to 33% longer than the distance between two genes (McBride and Athwal, 1976, 1977, 1979; Ruddle and McBride, 1977; McBride et al., 1978; Willecke et al., 1976c). The region coding for both *tk* and *galk* is estimated to be no longer than 0.2% of the human haploid genome. Thus, the transgenome must be no longer than approximately 0.25% of the haploid genome, or about 7×10^6 nucleotide pairs (McBride and Athwal, 1976; Ruddle and McBride, 1977; McBride et al., 1978; Willecke, 1978).

The assessment of transgenome size in terms of absolute amounts of DNA being transferred is possible by nucleic acid hybridization techniques. The current estimates obtained with these methods indicate that the fragment is sometimes as large as 0.2% to 0.8% of the haploid human genome (Olsen, McBride, and Aulakh, unpublished results). However, the possibility of multiple fragments accounting for this size has not been excluded yet.

Transfer of an Intact Chromosome or Large Chromosome Fragments

A large portion of the human X chromosome has been shown to be transferred to mouse A_9 cells when donor chromosomes were coated with lipid (Mukerjee et al., 1978). Although, three syntenic X-linked markers were cotransferred by this procedure, no cytologically detectable donor chromosomes were observed. In another type of procedure, an intact human X chromosome was transferred to an HPRT-deficient segregated hybrid (Chinese hamster $\times$ human) recipient cell line containing a few human chromosomes (Wullems et al., 1976). In both of these reports, the experimental design does not exclude other possible interpretations (McBride and Athwal, 1979).

Miller and Ruddle (1978) reported the transfer of cytologically detectable donor-chromosome fragments by modifying the original chromosome-transfer method. An interspecific gene-transfer system was used that involved the incubation of human chromosomes with mouse A_9 cells and selecting for *hprt*-gene transfer in HAT-selective medium. The new modified method differed from the usual technique (McBride and Ozer, 1973a, 1973b) with respect to the method of incubation of chromosomes with cells and in the treatment of the cells before and after the incubation period. Monolayers of recipient A_9 cells were pretreated with a mixture of colchicine (0.75 μg/ml), colcemid (0.75 μg/ml), and cytochalasin D (2 μg/ml) for a 5 minute interval immediately before adding the precipitated chromosomes in 250 mM $CaCl_2$ to the cell monolayer and incubating for 30 minutes at room temperature. The chromosome suspension was diluted to one-tenth with complete medium, and the cells were further incubated at 37°C for 4.25 hours. The monolayers were then treated with 10% dimethyl sulfoxide for 30 minutes. After replacement of the medium and incubation at 37°C for 18 hours, the cells were trypsinized and plated under cloning conditions in HAT selective medium. Using these procedures, the frequency of *hprt* transfer was increased about 100 times (10^{-5}) and X-linked syntenic markers (HPRT, G6PD, and PGK) were transferred in various combinations. Cytological analysis of these gene-transfer clones sometimes revealed a human chromosome fragment or translocation of a human chromosome piece onto a mouse chromosome. Although no intact human chromosome could be detected, there was some correlation between the size of the donor fragment (transgenome) and the number of syntenic markers observed in different clones. Only a single donor fragment was detected in any transformant, and none of the 25 clones expressed the donor form of any of four different autosomal markers.

The mechanism by which both calcium phosphate and dimethyl sulfoxide treatment increases the size and frequency of functional gene transfer is not clear. However, dimethyl sulfoxide is known to increase the polyethylene glycol-induced fusion of;cell membranes (Norwood et al., 1976), and it may also affect DNA metabolism (Miller and Ruddle, 1978). The calcium phosphate-precipitated chromosomes also may be less accessible to nucleolytic cleavage. This new improvement in CMGT technology provides a method to isolate at high frequency transformants with transgenomes of variable sizes. This should allow genetic mapping of both closely linked and syntenic genes that are separated by a considerable distance.

A recent report (Klobutcher and Ruddle, 1979) suggests that very large, cytologically detectable chromosome fragments may even be transferred by the original CMGT method (McBride and Ozer, 1973a, 1973b), albeit at a lower frequency than that observed with the modified procedure (Miller and Ruddle, 1978). Klobutcher and Ruddle (1979) used an interspecific transfer system involving human HeLa chromosomes and mouse *tk*$^-$ recipient cells with HAT selection to detect *tk* transfer. Transformants (*tk*$^+$) were isolated at the anticipated low frequency of 1-5 $\times$ 10^{-7}, and cotransfer of the tightly linked markers galactokinase *(galk)* and procollagen I *(PCI)* was observed in two out of seven and one out of seven clones, respectively. Moreover, cytologically detectable chromosome fragments representing as much as 1%, or more, of the human haploid genome were also found in two of these clones. Concordant segregation of the cytological fragment with the *tk* marker was observed in both cases. All seven transformants were also examined for expression of the chromosomal donor form of 20 different asyntenic markers. One cell line expressed the donor form of adenylate kinase, which has been assigned to human chromosome 9, and another transformant expressed human mannose phosphate isomerase, a human chromosome-15 marker. Moreover, these asyntenic markers segregated concordantly with human thymidine kinase, indicating a close association of each of these asyntenic genes with the selected human thymidine kinase marker.

The findings of Klobutcher and Ruddle (1979) clearly differ from all previous reports in several respects. Cytologically detectable donor-chromosome fragments had never previously been detected after chromosome transfer by the original method (McBride and Ozer, 1973a; Burch and McBride, 1975; Willecke and Ruddle, 1975; Wullems et al., 1975; Willecke et al., 1976c.; Athwal and McBride, 1977; McBride et al., 1978). This may reflect the relatively small number of transformant cell lines that have been previously examined and the fact that not all lines were examined cytologically at the earliest possible interval after transfer. However, this discrepancy most probably results from technical improvements in staining procedures, and it indicates a markedly enhanced sensitivity of the Giemsa 11 staining method for detecting even small human donor-chromosome fragments in a mouse background.

The prior failure to detect the transfer of asyntenic markers (Willecke and Ruddle, 1975; Willecke et al., 1976c.; McBride et al., 1978; Miller and Ruddle, 1978) is not readily explicable. Klobutcher and Ruddle (1979) point out that their recent results could be explained by rearrangements involving chromosomes 17, 9, and 15 in the HeLa donor line, thereby creating artificial synteny groups that were subsequently picked up in the transfer experiments. They favored the alternative explanation that the cotransfer of asyntenic genes arose through the uptake of more than a single chromosome by each of these two transformants. It should be noted that nonselective cotransfer of any specific marker was detected with a frequency of about 1% (that is, 2 markers detected among 20 examined times 7 cell lines, or 2/140), whereas the transfer frequency of the selected marker was about 10^{-7}. Thus, the latter alternative is plausible only if there is a very small fraction (that is, about 10^{-5}) of the total recipient cells that are competent for functional chromosome transfer. Moreover, the observed concordant segregation of asyntenic genes remains puzzling. A possible explanation could involve a much higher frequency for intraspecies recombination involving multiple-donor fragments than for interspecies recombination between a donor fragment and the chromosomal DNA of the recipient cell.

Unstable Expression of Transgenome

Gene-transfer clones can be classified as either stable or unstable, depending upon their ability to retain the transgenome in the absence of selective pressure (McBride and Ozer, 1973a; Degnen et al., 1975). Unstable clones become stabilized following continuous growth in either selective or nonselective media. In an unstable clone, the frequency of transgenome loss has been calculated to be 10^{-1} per cell per generation as compared to a low frequency (10^{-5}/cell/generation) of stabilization (Degnen et al., 1975). Stabilization of the transgenome is a unidirectional process; that is, the unstable transgenote can become stabilized, whereas the reverse is not observed. Stable subclones can thus be produced from any unstable clone following its continuous cultivation in selective or nonselective media (Degnen et al., 1976; Willecke et al., 1976c).

There is now both biochemical (Burch and McBride, 1975; Willecke and Ruddle, 1975; Wullems et al., 1975) and cytological (Miller and Ruddle, 1978; Klobutcher and Ruddle, 1979) evidence indicating that the transgenome exists as an acentric fragment in cells exhibiting the unstable phenotype. Thus, a cell frequently might not receive a copy of the transgenome at cell division since there is no mechanism for Mendelian segregation of an acentric fragment. A potential mechanism for survival of unstable transformants could be accumulation of multiple copies of the transgenome in one progeny cell through successive replications and cell divisions until there was a high probability of each daughter cell receiving at least one copy of the transgenome by random segregation. This possibility was first suggested by a report (Degnen et al., 1977) indicating that the specific activity of a transgenome-coded enzyme (HPRT) was 7 to 58 times higher in extracts of unstable transformants than in those of the donor cells or stable transformants. This observation has been confirmed in another laboratory (Scangos et al., 1979).

A recent report (Scangos et al., 1979) that multiple copies of a transferred gene can be directly detected in unstable transformants greatly strengthens this hypothesis. Metaphase chromosomes were isolated from a cell line carrying an integrated copy of the thymidine kinase gene of Herpes simplex virus (HSV) type I and incubated with mouse LMTK$^-$ cells. Four independent unstable transformants expressing HSV-thymidine kinase activity were isolated in HAT-selective medium. The DNA from each of these transformants and their stable derivatives was analyzed by filter hybridization of restriction fragments with a radiochemically labeled HSV *tk* probe after agarose-gel electrophoresis. Each of the unstable transformants contained a higher number of HSV *tk* copies than was found in their stabilized derivatives, but quantification of the copy number was not possible. The analysis was further complicated by the fact that the donor line probably also contained multiple copies of the HSV *tk* gene, although the evidence indicated that this resulted from an increased copy number of the entire chromosome bearing the integrated HSV *tk* gene. Supporting the conclusion that multiple copies of the HSV *tk* gene exists in unstable transformants, Scangos and his colleagues found an elevated specific activity of the HSV *tk* enzyme in these cell lines compared to the stablized derivatives. These observations are all consistent with survival of unstable transformants by random segregation of the transgenome at cell division. The possibility that an elevated copy number of HSV *tk* in unstable transformants was a consequence of the uptake of multiple *tk* genes, on either one or several chromosome fragments, cannot be excluded, however.

Stabilization of the Transgenome by Integration

Stabilization of the transgenome represents its physical integration into the chromosomal DNA of the recipient cell. Integration of an acentric chromosome fragment into a recipient chromosome permits its orderly distribution into daughter cells and thus confers stability to the transgenote.

Evidence for integration of the transgenome was first provided by sequential transfer of human *hprt* and *tk* genes following two successive cycles of chromosome isolation (Athwal and McBride, 1976, 1977; Willecke et al., 1976a, 1976b). These experiments were based on the hypothesis that integration of the transgenome into the recipient genome should permit isolation of chromosomes from the transgenote and serial transfer of the genetic marker to another recipient cell line. The human *hprt* gene was first transferred to mouse A_9 cells (Burch and McBride, 1975). Metaphase chromosomes were then isolated from the mouse cells stably expressing human *hprt* and incubated with Chinese hamster HPRT-deficient cells. A number of colonies expressing this human genetic marker were isolated following serial transfer. Similar results were reported for the serial transfer of the human *tk* gene (Willecke et al., 1976a, 1976b).

Definitive evidence for integration of the transgenome into nonhomologous chromosome sites was furnished by the genetic analysis of transgenotes by microcell fusion (Fournier and Ruddle, 1978a, 1978b). This procedure provides a method to transfer an individual chromosome, or a few intact chromosomes, to a recipient cell (Ege and Ringertz, 1974; Fournier and Ruddle, 1977a). Thus, hybrid cell lines carrying one, or a few, donor chromosomes can be isolated for immediate genetic analysis (described later). Microcells were prepared from a stable transgenote population of mouse A_9 cells expressing human *hprt* and fused with HPRT-deficient Chinese hamster cells (Fournier and Ruddle, 1977a, 1977b). Three independent hybrid clones isolated in HAT medium were subjected to biochemical and cytological analysis. The transferred human *hprt* was mapped to a different mouse chromosome in each instance. These three chromosomes included mouse chromosome 14, chromosome 14 with centric fusion to chromosome 15, and an unidentified mouse chromosome fragment. Neither a mouse X chromosome nor mouse X-chromosomal markers were ever detected. These results were construed to indicate multiple integration sites for the transferred genes. However, in two of the three cases, the transgenome was mapped to chromosome 14. These interpretations assume that rearrangements of mouse recipient chromosomes occurred prior to the integration of the transgenome. In experiments using microcells prepared from unstable transgenotes, Fournier and Ruddle were unable to isolate hybrids.

Willecke et al. (1978) reported similar results using microcells prepared from two independent stable mouse transgenotes expressing human *tk* that were fused with Chinese hamster cells. In one of the three isolated hybrids, integration of human *tk* was mapped to mouse chromosome 9, whereas the normal mouse *tk* gene is located on chromosome 11. In the other two hybrids, no mouse chromosome could be detected, and these hybrids expressed a high degree of stability of the human *tk* on back selection. It was postulated that the human *tk* gene was probably translocated onto a Chinese hamster chromosome in these two hybrids.

The microcell-fusion approach has also been used in our laboratory (Athwal and McBride, unpublished results) to analyze integration sites of the transgenome. A transgenote resulting from sequential transfer of the human *hprt* gene to mouse cells and then to Chinese hamster cells was stabilized and subcloned. Single-hybrid clones were isolated in HAT and analyzed after fusion of mouse HPRT-deficient TG8 cells with microcells prepared from each of two subclones. The human *hprt* gene was assigned to a small acrocentric hamster chromosome in one case, and it was present on a very large Chinese hamster metacentric chromosome in the other instance (McBride and Athwal, 1979). Subclones from both of these hybrids and their back-selected populations were analyzed biochemically to determine the Chinese hamster markers that were linked to the human *hprt* gene. The results indicated that, in both hybrids, human *hprt* was syntenic with the same Chinese hamster genes, that is, phosphoglucomutase-2 (PGM-2), 6-phosphogluconate dehydrogenase (PGD), adenylate kinase-2 (AK-2), and enolase (ENO-1) (Athwal et al., 1979). Thus, the two hybrids cytologically seemed to exhibit two different-sized hamster chromosomes, but biochemical analysis showed that the putative integration site of the transgenome may be the same in both subclones of this transgenote. This is likely since stabilization occurs at a low frequency of about 10^{-5} /cell/generation (Degnen et al., 1975) and confers a selective growth advantage in selective media. Hence, the progeny of the earliest integration event could be expected to predominate in a mass culture. The observed cytological differences in these two hybrids presumably represent chromosomal rearrangements in the transgenote following integration of the transgenome. An earlier report (Fournier and Ruddle, 1977b) is also consistent with this interpretation, and more recent evidence (Klobutcher and Ruddle, 1979) indicates that it is correct.

The recent report by Klobutcher and Ruddle (1979) provides further information relating to the integration of the transgenome. They studied the integration of a large human chromosome fragment containing the genes for thymidine kinase (TK), galactokinase (GalK), and type I procollagen (PCI) into the chromosomal DNA of mouse recipient cells using cytological and biochemical techniques. Nine independent subclones were derived from this single unstable transgenote and analyzed after stabilization. The human fragment was integrated into a single mouse chromosome in each case, but the mouse chromosome involved differed between these subclones. Stabilization in each instance was accompanied by acquisition of a mouse centromere plus various lengths of a mouse chromosome. The length of the human fragment always decreased following translocation and was correlated with the number of human genes retained, thereby permitting detailed localization and ordering of these three human genes. Moreover, it was reported that shortening of the fragment always proceeded from the centromeric end. It was proposed that deletion mapping following stabilization of visible chromosome fragments could provide a means of deriving qualitative intergenic mapping data.

Microcell-Mediated Gene Transfer

Another recently developed method to transfer a limited amount of genetic information from one cell to another is microcell hybridization. This technique offers the possibility of transferring one, or a few, donor chromosomes to a recipient cell (Ege and Ringertz, 1974; Fournier and Ruddle, 1977a). The induction of abnormal mitosis in mammalian cells inhibits the formation of an intact nucleus, and instead a number of subdiploid nuclei (micronuclei) are formed. When experimentally isolated, these consist of a subdiploid nucleus surrounded by a rim of cytoplasm and a plasma membrane and are called microcells (Ege and Ringertz, 1974). Microcells are not capable of self-multiplication but can be fused with intact cells. Since each microcell contains only a fraction of the donor genome, the isolated hybrids carry only a limited number of chromosomes derived from the microcell donor. The microcell-transfer technique thus utilizes the segregation of chromosomes within one parental cell prior to fusion with whole cells, and it permits immediate analysis of the resultant hybrids.

The preparation of microcells involves two distinct processes. The first step involves the induction of micronucleation in a population of cells, and the second step is the isolation of resultant individual micronuclei. In normal cell division, chromosomes of a cell are duplicated at mitosis and are subsequently separated to form two nuclei on reorganization of the nuclear envelope. Cytokinesis occurs synchronous with nuclear reorganization, and the cell divides to form two daughter cells. Interference with the normal process of mitosis or chromosome separation has been found to result in the formation of multiple nuclei without cytokinesis, each containing a subset of the cell genome (Ege and Ringertz, 1974). Micronuclei formation can be induced by a number of agents that interfere with the assembly of microtubes and mitotic spindle formation. The prevention of microtubule assembly by mitotic inhibitors results in a divergent anaphase movement of chromosomes. The scattered individual chromosomes or small groups of chromosomes function as the sites for the reassembly of nuclear membranes and, thus, form micronuclei (Ege et al., 1977; Phillips and Phillips, 1969). The best-known agents that induce micronucleation are colchicine, colcemid, vinblastine, and X rays (Ege et al., 1977). Colchicine, colcemid, and vinblastine have been found to give approximately the same degree of micronucleation (Ege et al., 1977). The percent of micronucleated cells and the number of micronuclei in individual cells varies, depending upon the concentration of the mitotic inhibitor, length of exposure, and cell line used. It may also depend on the growth rate of the cells. Cells of different species of origin vary in their sensitivity to mitotic inhibitors (Athwal, unpublished results). In a number of Chinese hamster and mouse cell lines, it is possible to induce micronucleation in 70% to 90% of the cells (Ege et al., 1977). The percentage of micronucleated cells and the relative number of micronuclei per cell increase with a greater interval of treatment with mitotic inhibitor. Colcemid treatment for four days resulted in a higher fraction of micronucleated cells than was obtained after two days (Fournier and Ruddle, 1977a, 1977b). A synchronized population of mouse cells, prior to the addition of colchicine, required a shorter time interval for the induction of micronucleation (Athwal and McBride, unpublished results). Time of exposure to mitotic inhibitors is an important variable since the frequency of viable hybrids following microcell fusion was higher when microcells from shorter exposure were used (Fournier and Ruddle, 1977a). Photomicrographs of appropriately stained preparations of a micronucleated cell population (figure 1) reveal that micronuclei vary both in size in a given cell and in number between cells. A rough correlation is found to exist between the modal chromosome number of the cell line and the maximum number of micronuclei in the cells. Thus, the maximum number of micronuclei observed in Chinese hamster cells (11 pairs of chromosomes) is very close to 22 (Ege et al., 1977; Phillips and Phillips, 1969).

Preparation of Microcells

Microcells can be isolated from the micronucleated cells by

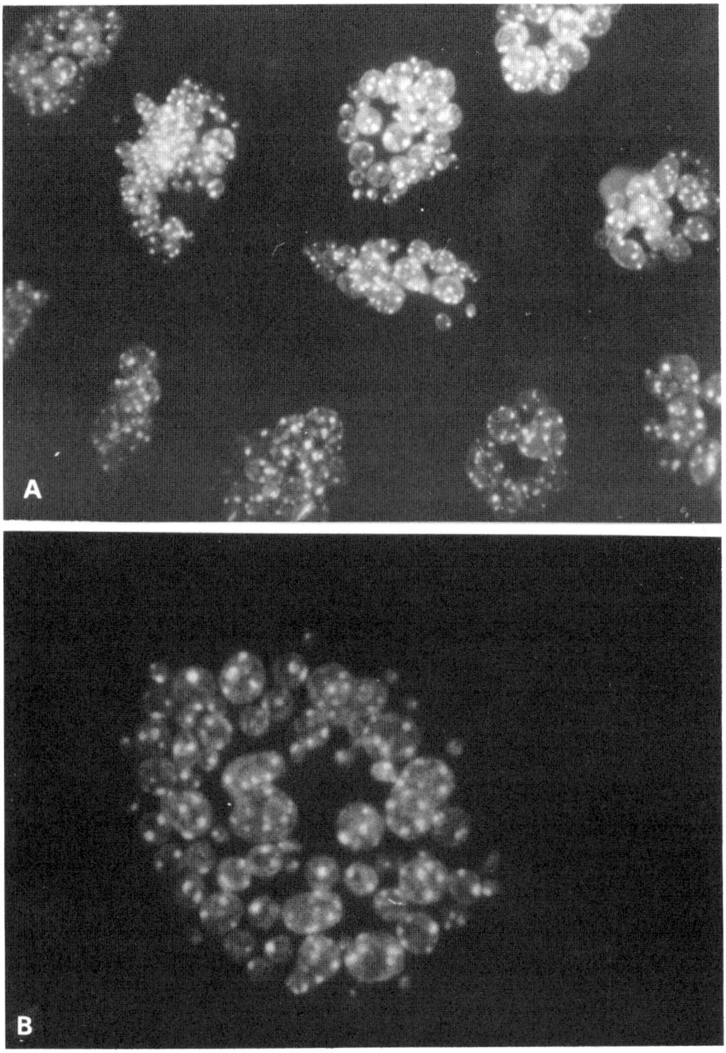

Fig. 1. Photomicrographs of mouse L cells stained with Hoechst 33258, showing flourescent micronuclei. (A) Group of micronucleated cells showing brightly flourescent chromocenters; each chromocenter represents a centromere and indicates a single chromosome. (B) Higher magnification of an individual micronucleated cell demonstrating variation in the number of chromocenters in different micronuclei. None of the micronuclei in this cell contain more than five chromocenters, and many appear to contain a single chromosome.

centrifugation either attached to a solid surface or in a discontinuous gradient of ficoll (Ege and Ringertz, 1974; Ege et al., 1977; Fournier and Ruddle, 1977a; Athwal et al., 1979) after exposure to cytochalasin B. Cells are grown on plastic coverslips punched out at the bottom of tissue culture-plates. Following the induction of micronucleation, these coverslips are removed from media and inserted, facing downward, into centrifuge tubes containing buffered saline. The position of a coverslip is fixed with an inserted solid plug so that it remains at the top of the saline. The coverslips are then centrifuged at 12,000 × g for 10 minutes in a prewarmed (37 °C) centrifuge. This centrifugation removes loosely attached cells that would otherwise contaminate the microcell population. Following the first centrifugation, the coverslips are transferred into clean tubes containing cytochasin B (10 μg/ml) in saline and centrifuged at 25,000 × g for 20 minutes. Prior to centrifugation, the media, rotor, and centrifuge are warmed to 37 °C. The microcells are pulled out of the cells into the pellet by the centrifugal force, whereas cytoplasts stay attached to the coverslips.

Large quantities of microcells can also be prepared by zonal sedimentation of micronucleated cells in discontinuous ficoll gradients containing cytochalasin B (Athwal et al., 1979; Wigler and Weinstein, 1975). Micronucleated cells are harvested from monolayers or suspension and are incubated in complete media containing cytochasin B (10 μg/ml) at 37 °C for 15 minutes. After the pretreatment, the cell suspension (1-2 × 10^7 cells/ml) is made 12% ficoll containing 10 μg/ml of cytochasin B. This cell suspension is then layered over previously formed ficoll gradients as described. A 20% solution of ficoll (5 ml) is placed in a polycarbonate tube (SW 27 rotor) and overlayered with 15 ml of 16% ficoll. The cells are centrifuged at 100,000 × g for 90 minutes in a swinging bucket rotor at 37 °C. All gradients, rotor, and centrifuge chamber are warmed to 37 °C prior to centrifugation. Microcells are collected in the pellet and at the 16% to 20% ficoll interface. The cytoplasts band at the 12% to 16% ficoll interface, and cellular debris from broken cells remains on top of the gradient.

A different method of producing microcells in human cells

(HeLa) has also been reported (Johnson et al., 1975; Shor et al., 1975). Synchronized cells were arrested in mitosis by exposure to nitrous oxide (5 atmospheres) before incubation at cold temperatures (5 °C) for nine hours. Aberrant chromosome segregation and cytokinesis were observed when these cells were transferred to 37 °C after the cold treatment. The dividing cells were found to form structures described as "bunches of grapes." These structures consisted of cytoplasmic fragments containing one or several micronuclei detached spontaneously from the main body of the cell. These detached fragments have been termed *minisegregants* (Johnson et al., 1975; Schor et al., 1975; Tourien et al., 1978). Similarly, Chinese hamster V79 cells blocked with colchicine (20 μg/ml) for 36-48 hours form a large number of budlike protuberances that spontaneously pinch off to produce microcells (Athwal and McBride, unpublished results). The separation of these spontaneously formed microcells can be facilitated by zonal sedimentation on ficoll gradients.

A crude microcell preparation is a very heterogeneous population of various-sized classes of microcells and nuclei (minicells). The population of microcells can be fractionated into different size classes by unit-gravity sedimentation on ficoll gradients (Schor et al., 1975) or on bovine serum albumin gradients (Fournier and Ruddle, 1977a, 1977b). Microcells of various sizes may also be fractionated by filtration through nuclepore filters of pore diameter from 1 to 5 μm (Ege et al., 1977).

The DNA content of individual isolated microcells can be assessed by flow microfluorometry (Athwal and McBride, unpublished results) after staining with fluorochromes such as mithramycin or by microspectrophotometry (Sekiguchi et al., 1978). The DNA content of rat kangaroo microcells varies from approximately the value expected for a single chromosome up to that of the entire genome. DNA measurements on individual micronuclei following detergent isolation (Goto and Ringertz, 1974) showed a wide variation. Some nuclei were found to contain over four times the DNA content of a G1 cell. However, approximately 35% of the total nuclei population contained less than the normal G1 DNA content. The smallest micronuclei in this group had a DNA content approaching that of individual Chinese hamster chromosomes. The number of chromosomes in individual micronuclei of murine cells can also be determined by counting the number of brightly fluorescent chromocenters in photomicrographs of these particles after staining with Hoechst 33258. The results of such a study indicated a correlation between the size of the micronuclei and the number of chromocenters (McBride and Athwal, 1979). The smallest micronuclei had only one highly stained centromeric region, suggesting that they contained only a single chromosome (figure 1B).

The microcells prepared by various methods can be fused with whole cells of the same or different species using inactivated Sendai virus or polyethylene glycol (Fournier and Ruddle, 1977a; Tourien et al., 1978; Athwal et al., 1979; Schor et al., 1975). The microcell-hybridization method permits the transfer of a relatively controlled number of intact chromosomes to another cell. By using the smallest class of microcells, it is possible to introduce

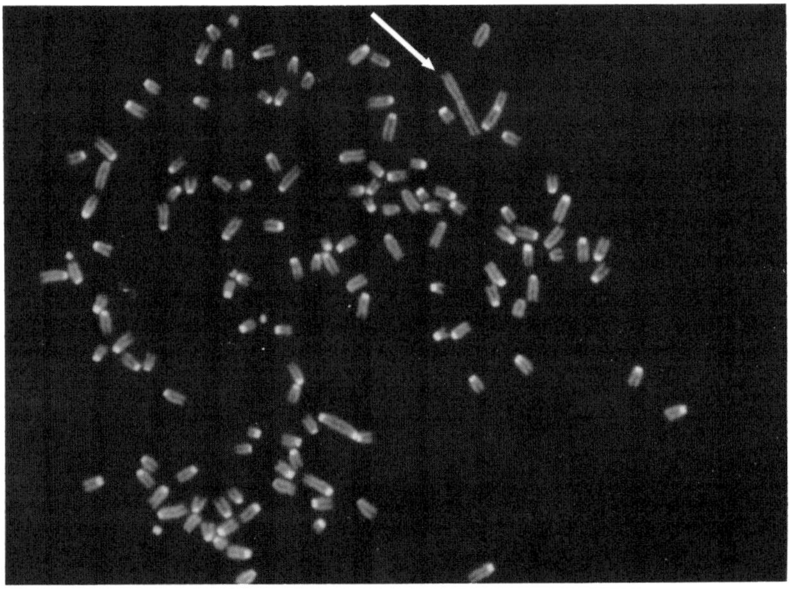

Fig. 2. Metaphase spread of a microcell hybrid between Chinese hamster microcells and mouse cells, stained with Hoechst 33258. A single Chinese hamster chromosome is present in the mouse cell.

a single chromosome, or a very small number of chromosomes, into recipient cells.

Fournier and Ruddle (1977a, 1977b) first used this technique for genetic analysis of somatic cells. Microcells were prepared from a mouse cell line expressing human *hprt* (transferred by CMGT) and fused with Chinese hamster *hprt*$^-$ cells. Three independent hybrids isolated in HAT medium expressed human *hprt* and contained one to four human chromosomes. The transferred human *hprt* gene was assigned to mouse chromosome 14 following biochemical and cytological analysis in two of these hybrids. Using the same general procedure, integration of the human *tk* gene has been mapped to mouse chromosome 9 in a clone isolated by CMGT (Willecke et al., 1978).

Subsequently, this method was used to transfer Chinese hamster chromosomes into mouse cells. In our laboratory, we prepared microcells from Chinese hamster V79 cells expressing transferred human *hprt*. The microcell population was fractionated by unit-gravity sedimentation, and the smallest class of microcells was fused with mouse TG8 cells. Cytological analysis of two hybrid clones isolated in HAT medium revealed that only a single Chinese hamster chromosome was transferred through this method (figure 2). The presence of a single chromosome in these clones was further confirmed on biochemical analysis (Athwal et al., 1979). A limited number of human chromosomes have also been transferred into mouse A_9 cells by hybridizing the size-fractionated minisegregants (Tourien et al., 1978).

Microcell hybridization thus permits the transfer of single, or a limited number of, chromosomes of one cell into another. The number of input chromosomes can be controlled by using different-sized classes of microcells after fractionation (Fournier and Ruddle, 1977; Athwal et al., 1979). Although it was reported (Fournier and Ruddle, 1977a, 1977b; Tourien et al., 1978) that chromosome segregation does occur in these microcell hybrids, it should be both easier and quicker to achieve a stable hybrid karyotype. Finally, a major advantage of this technique is the ability to analyze segregants of a parental cell type in which segregation cannot be achieved by conventional somatic cell hybridization.

The limitations of this method relate to the fact that most chromosomes do not carry loci for phenotypic markers that are subject to selection. Thus, it is difficult to isolate transfer clones for every chromosome at the present time. However, some increase in the number of selection systems can be anticipated in the future. Moreover, as already discussed, the ability to integrate selectable markers into multiple integration sites by chromosome-mediated gene transfer would permit the introduction of a selectable marker into any chromosome.

Conclusion

The ability to produce cells containing different proportions of a transferred genome would undoubtedly permit a more rapid analysis of eukaryotic genomes. Recently developed techniques offer the possibility for the transfer of genetic material varying from a single gene to the entire genome from one mammalian cell to another.

Chromosome-mediated gene transfer provides a method for transferring genetic material varying from a single functional gene to a cytologically recognizable portion of a chromosome. This method is ideally suited for fine structural mapping of eukaryotic chromosomes. The expanded range of transgenome sizes now available, by application of the calcium phosphate transfer method and dimethyl sulfoxide treatment, allows the analysis of syntenic but distantly linked genes as well. This method may also be very useful for the analysis of malignant transformation of cells. Tumorigenicity has been reported to be phenotypically dominant (Spandidos and Siminovitch, 1977, 1978; Cassingena et al., 1977) as determined by CMGT.

A controlled number of intact chromosomes can be transferred by microcell fusion. Moreover, the number of transferred chromosomes can be varied by size fractionation of microcells prior to hybridization. This method also permits control over the direction of chromosome segregation.

Since the integration of genes transferred by CMGT is not locus specific, it is theoretically possible to integrate a selectable marker into any recipient chromosome. This could be especially useful if combined with microcell hybridization to transfer the chromosome carrying the selectable marker. Thus, a panel of cell lines could be developed in which a single specific chromosome of one parental type is stably maintained under selective pressure.

The availability of cell lines carrying various numbers of transferred genes would serve a vast number of purposes in the analysis of the eukaryotic genome. A few of these include the identification of loci for many antigens and the ability to produce useful immunological reagents. These cell lines could be used to isolate defined portions of donor chromosomal DNA by various available methods. This material could also serve many interesting purposes, including the cloning of these genes using recombinant-DNA technology and investigating sequence organization of the eukaryotic genome.

These techniques can also be extended to modify the plant genomes in cell cultures. Plant protoplasts are known to take up a variety of organelles in suspension, such as chloroplasts and nuclei (Potrykus and Horz, 1976; Potrykus and Hoffman, 1973; Cocking, 1977; Lörz and Potrykus, 1978) and can also be fused to generate hybrid protoplasts combining genomes of different species (Carlson, 1973; Gamborg, 1977; Melchers et al., 1978; Cocking, 1977). Under suitable culture condition, plant protoplasts can generate a new cell wall and form a callus that can eventually develop into a mature plant (other papers in this volume). Thus, this would allow the incorporation of desired genes by these methods for the practical purpose of improving our food plants. The methods that transfer a limited number of genes might also avoid the concomitant transfer of genes that are undesirable for the economic cultivation of crop plants.

References

Athwal, R. S., and McBride, O. W. (1977) Serial transfer of a human gene to rodent cells by sequential chromosome mediated gene transfer. *Proc. Nat. Acad. Sci., U.S.A. 74,* 2943-47.

Athwal, R. S.; Gregorio, T.; Minna, J. D.; and McBride, O. W. (1979) Application of gene transfer techniques for genetic analysis of Chinese hamster cells. *Fed. Proc. 38,* 783.

Berg, T., Melbye, B.; Johnsen, R.; and Prydz, H. (1975) Activity of lysosomal enzymes in the various cell cycle phases of synchronized HeLa cells. *Exp. Cell 94,* 106-10.

Burch, J. W., and McBride, O. W. (1975) Human gene expression in rodent cells after uptake of isolated mataphase chromosomes. *Proc. Nat. Acad. Sci., U.S.A. 72,* 1797-1801.

Carlson, P. S. (1973) The use of protoplasts for genetic research. *Proc. Nat. Acad. Sci., U.S.A. 70,* 598-602.

Cassingena, R.; Suarez, H. G.; and Lavialle, C. (1977) Transformation cellulaire *in vitro* par le transfert de chromosomes metaphasiques isoles. *C. R. Acad. Sci. Paris 285,* 1603-6.

Cassingena, R.; Suarez, H. G.; Lavialle, C.; Persuy, M. A.; and Ermonval, M. (1978) Transformation of normal diploid cells by isolated metaphase chromosomes of virus-transformed or spontaneous tumor cells. *Gene 4,* 337-49.

Cocking, E. C. (1977a) Plant protoplast fusion: Progress and prospects for agriculture. *Recombinant Molecules: Impact on Science and Society* (R. F. Beers, Jr., and E. C. Bassett, eds.), pp. 195-208. Raven Press, New York.

Cocking, E. C. (1977b) Uptake of foreign genetic material by plant protoplasts. *Int. Rev. Cytol.* 48; 323-43.

Degnen, G. E.; Miller, I. L.; Eisenstadt, J. M.; and Adelberg, E. A. (1976) Chromosome mediated gene transfer between closely related strains of cultured mouse cells. *Proc. Nat. Acad. Sci., U.S.A. 73,* 2838-42.

Degnen, G. E.; Miller, I. L.; Adelberg, E. A.; and Eisenstadt, J. M. (1977) Overexpression of an unstably inherited gene in cultured mouse cells. *Proc. Nat. Acad. Sci., U.S.A. 74,* 3956-59.

Ege, T., and Ringertz, N. R. (1974) Preparation of microcells by enucleation of micronucleate cells. *Exp. Cell Res. 87,* 378-82.

Ege, T.; Ringertz, N. R., Hamberg, H.; and Sidebottom, E. (1977) Preparation of microcells. In *Methods in Cell Biology* (D. M. Prescott, ed.), volume 15, pp. 339-57. Academic Press, New York.

Elsevier, S.; Kucherlapati, R. S.; Nichols, E. A.; Creagan, R. P.; Giles, R. E.; Ruddle, F. H.; Willecke, K.; and McDougall, J. (1974) Assignment of the gene for galactokinase to human chromosome 17 and its regional localisation to band q21-22. *Nature 251,* 633-36.

Fournier, R. E. K., and Ruddle, F. H. (1977a) Microcell-mediated transfer of murine chromosomes into mouse, Chinese hamster, and human somatic cells. *Proc. Nat. Acad. Sci., U.S.A. 74,* 319-23.

Fournier, R. E. K., and Ruddle, F. H. (1977b) Stable association of the human transgenome and host murine chromosomes demonstrated with trispecific microcell hybrids. *Proc. Nat. Acad. Sci., U.S.A. 74,* 3837-41.

Gamborg, O. L. (1977) Somatic cell hybridization by protoplast fusion and morphogenesis. In *Plant Tissue Culture and Its Bio-technological Application* (W. Barz, E. Reinhard, and M. H. Zenk, eds.), pp. 287-301. Springer, Berlin.

Goto, S., and Ringertz, N. R. (1974) Preparation and characterization of chick erythrocyte nuclei from heterokaryons. *Exp. Cell Res. 85,* 173-81.

Johnson, R. T.; Mullinger, A. M.; and Skaer, R. J. (1975) Perturbation of mammalian cell division; human mini segregants derived from mototic cells. *Proc. R. Soc. London, Ser. B 189,* 591-602.

Klobutcher, L. A., and Ruddle, F. H. (1979) Phenotype stablization and integration of transferred material in chromosome-mediated gene transfer. *Nature 280,* 657-60.

Kozak, C. A., and Ruddle, F. H. (1977) Assignment of the genes for thymidine kinase and galactokinase to Mus musculus chromosome 11 and the preferential segregation of this chromosome in Chinese hamster/mouse somatic cell hybrids. *Somatic Cell Genet. 3,* 121-33.

Littlefield, J. W. (1964) Selection of hybrids from matings of fibroblasts *in vitro* and their presumed recombinants. *Science 145,* 709-10.

Lörz, H., and Potrykus, I. (1978) Investigations on the transfer of isolated nuclei into plant protoplasts. *Theor. Appl. Genet. 53,* 251-56.

McBreen, P.; Oskwiszewski, K. G.; Chern, C. J.; Melman, W. J.; and Croce, C. M. (1977) Synteny of the genes for thymidine kinase and galactokinase in the mouse and their assignment to mouse chromosome 11. *Cytogenet. Cell Genet. 19,* 7-13.

McBride, O. W. (1974) Metaphase chromosome uptake by mammalian cells and expression of the genes transferred. In *Stadler Genetics Symposia* (G. P. Redei and G. Kimber, eds.), volume 6, pp. 53-75. University of Missouri Press, Columbia, Missouri.

McBride, O. W., and Athwal, R. S. (1976) Genetic analysis by chromosome-mediated gene transfer. *In Vitro 12,* 777-86.

McBride, O. W., and Athwal, R. S. (1977) Chromosome-mediated gene transfer with resultant expression and integration of the transferred genes in eukaryotic cells. *Brookhaven Symp. Biol. 29,* 116-26.

McBride, O. W., and Athwal, R. S. (1979) Genetic analysis of eukaryotic cells by chromosome-mediated gene transfer and other methods. In *Concepts of the Structure and Function of DNA, Chromatin, and Chromosomes* (A. S. Dion, ed.), pp. 229-63.

McBride, O. W.; Burch, J. W.; and Ruddle, F. H. (1978) Cotransfer of thymidine kinase and galactokinase genes by chromosome-mediated gene transfer. *Proc. Nat. Acad. Sci., U.S.A. 75,* 914-18.

McBride, O. W.; and Ozer, H. L. (1973a) Transfer of genetic information by purified metaphase chromosomes. *Proc. Nat. Acad. Sci., U.S.A. 70,* 1258-62.

McBride, O. W.; and Ozer, H. L. (1973b) Gene transfer with purified mammalian chromosomes. In *Possible Episomes in Eukaryotes: Le Petit Colloquia on Biology and Medicine* (L. G. Silvestri, ed.), volume 4, pp. 255-67. Amsterdam, North Holland.

McKusick, V. A.; and Ruddle, F. H. (1977) The status of gene mapping of the human chromosomes. *Science 196,* 390-405.

Maio, J. J., and Schildkraut, C. L. (1969) Isolated mammalian metaphase chromosomes II. Fractionated chromosomes of mouse and Chinese hamster cells. *J. Mol. Biol. 40,* 203-16.

Melchers, G.; Sacristan, M. D.; and Holden, A. A. (1978) Somatic hybrid plants of potato and tomato regenerated from fused protoplasts. *Carlsberg Res. Commun. 43,* 203-18.

Mendelsohn, J.; Moore, D. E.; and Salzman, N. P. (1968) Separation of isolated Chinese hamster metaphase chromosomes into three size-groups. *J. Mol. Biol. 32,* 101-12.

Miller, C. L.; and Ruddle, F. H. (1978) Co-transfer of human X-linked markers into murine somatic cells via isolated metaphase chromosomes. *Proc. Nat. Acad. Sci., U.S.A. 75,* 3346-50.

Mukherjee, A. B.; Orloff, S.; Butler, J. DeB.; Triche, T.; Lalley, P.; and Schulman, J. D. (1978) Entrapment of metaphase chromosomes into phospholipid vesicles (lipochromosomes): Carrier potential in gene transfer. *Proc. Nat. Acad. Sci., U.S.A. 75,* 1361-65.

Norwood, T. H.; Zeigler, C. J.; and Martin, G. M. (1976) Dimethyl sulfoxide enhances polyethylene glycol-mediated somatic cell fusion.

Peterson, J. L.; and McBride, O. W. (1980) Cotransfer of linked eukaryotic genes and efficient transfer of hypoxanthine phosphoribosyltransferase by DNA-mediated gene transfer. *Proc. Nat. Acad. Sci., U.S.A.* (in press).

Phillips, S. G., and Phillips, D. M. (1969) Sites of nucleolus production in cultured Chinese hamster cells. *J. Cell Biol. 40,* 248-68.

Potrykus, I., and Hoffman, F. (1973) Transplantation of nuclei into protoplasts of higher plants. *Z. Pflanzenphysiol. 69,* 287-89.

Potrykus, I., and Lörz, H. (1976) Organelle transfer into isolated protoplasts. In *Cell Genetics in Higher Plants* (D. Dudits, G. L. Farkas, P. Maliga, eds.), pp. 183-90. Akademiai Kiado, Budapest.

Ruddle, F. H., and Creagan, R. P. (1975) Parasexual approaches to genetics of man. *Ann. Rev. Genet. 9,* 405-86.

Ruddle, F. H., and McBride, O. W. (1977) New approaches to cell genetics: Cotransfer of linked genetic markers by chromosome mediated gene transfer. In *The Molecular Biology of the Mammalian Genetic Apparatus* (P. O. T'so, ed.), volume 2, pp. 163-69. North Holland Publishing Company, Amsterdam and New York.

Sabourin, D. J., and Davidson, R. L. (1979) Transfer of the Herpes simplex thymidine kinase gene from human cells to mouse cells by means of metaphase chromosomes. *Somat. Cell Genet. 5,* 159-74.

Scangos, G. A.; Huttner, K. M.; Silverstein, S.; and Ruddle, F. H. (1979) Molecular analysis of chromosome-mediated gene transfer. *Proc. Nat. Acad. Sci., U.S.A. 76,* 3987-90.

Schor, S. L.; Johnson, R. T.; and Mullinger, A. M. (1975) Perturbation of mammalian cell division. II: Studies on the isolation and characterization of human mini segregant cells. *J. Cell. Sci. 19,* 281-303.

Sekiguchi, T.; Shelton, K.; and Ringertz, N. R. (1978) DNA content of microcells prepared from rat kangaroo and mouse cells. *Exp. Cell Res. 113,* 247-58.

Shani, M.; Huberman, E.; Aloni, Y.; and Sachs, L. (1974) Activation of simian virus 40 by transfer of isolated chromosomes from transformed cells. *Virology 61,* 303-5.

Shani, M.; Aloni, Y.; Huberman, E.; and Sachs, L. (1976) Gene activation by transfer of isolated mammalian chromosomes. II: Activation of simian virus 40 by transfer of fractionated chromosomes from transformed cells. *Virology 70,* 201-5.

Spandidos, D. A., and Siminovitch, L. (1977a) Transfer of codominant markers by isolated metaphase chromosomes in Chinese hamster ovary cells. *Proc. Nat. Acad. Sci., U.S.A. 74,* 3480-84.

Spandidos, D. A., and Siminovitch, L. (1977b) Linkage of markers controlling consecutive biochemical steps in CHO cells as demonstrated by chromosome transfer. *Cell 12,* 235-42.

Spandidos, D. A., and Siminovitch, L. (1977c) Transfer of anchorage independence by isolated metaphase chromosomes in hamster cells. *Cell 12,* 675-82.

Spandidos, D. A., and Siminovitch, L. (1978) Transfer of the marker for morphologically transformed phenotype by isolated metaphase chromosomes in hamster cells. *Nature 271,* 259-61.

Szybalska, E. H., and Szybalski, W. (1962) Genetics of human cell lines. IV: DNA-mediated heritable transformation of a biochemical trait. *Proc. Nat. Acad. Sci., U.S.A. 48,* 2026-34.

Tourian, A.; Johnson, R. T.; Burg, K.; Nicolson, S. W.; and Sperling, K. (1978) Transfer of human chromosomes via human minisegregant cells into mouse cells and the quantitation of the expression of hypoxanthine phosphoribosyl transferase in the hybrids. *J. Cell Sci. 30,* 193-209.

Wigler, M.; Pellicer, A.; Silverstein, S.; and Axel, R. (1978) Biochemical transfer of single-copy eukaryotic genes using total cellular DNA as donor. *Cell 14,* 725-31.

Wigler, M. H., and Weinstein, B. I. (1975) A preparative method for obtaining enucleated mammalian cells. *Biochem. Biophys. Res. Commun. 63,* 669-74.

Willecke, K. (1978) Results and prospects of chromosomal gene transfer between cultured mammalian cells. *Theor. Appl. Genet. 52,* 97-104.

Willecke, K.; Davies, P. J.; and Reber, T. (1976a) Chromosomal gene transfer: Prospects of a new method for mapping closely linked human genes. *Cytogenet. Cell Genet. 16,* 405-8.

Willecke, K.; Davies, P. J.; and Reber, T. (1976b) Sequential cotransfer via purified metaphase chromosomes of two linked human genes into murine L-cells. *Hoppe Seyler's Z. Physiol. Chem. 357,* 341-42.

Willecke, K.; Lange, R.; Kruger, A.; and Reber, T. (1976c) Cotransfer of two linked human genes into cultured mouse cells. *Proc. Nat. Acad. Sci., U.S.A. 73,* 1274-78.

Willecke, K.; Mierau, R.; Kruger, A.; and Lange, R. (1978) Chromosomal gene transfer of human cytosol thymidine kinase into mouse cells. *Mol. Gen. Genet. 161,* 49-57.

Willecke, K., and Ruddle, F. H. (1975) Transfer of the human gene for hypoxanthine-guanine phosphoribosyltransferase via isolated human metaphase chromosomes into mouse L-cells. *Proc. Nat. Acad. Sci., U.S.A. 72,* 1792-96.

Wray, W. (1973) Isolation of metaphase chromosomes with high molecular weight DNA at pH 10.5. In *Methods in Cell Biology* (D. M. Prescott, ed.), volume 6, pp. 283-315. Academic Press, New York.

Wullems, G. J.; van der Horst, J.; and Bootsma, D. (1975) Incorporation of isolated chromosomes and induction of hypoxanthine phosphoribosyltransferase in Chinese hamster cells. *Somatic Cell Genet. 1,* 137-52.

Wullems, G. J.; van der Horst, J.; and Bootsma, D. (1976a) Expression of human hypoxanthine phosphoribosyltransferase in Chinese hamster cells treated with isolated human chromosomes. *Somatic Cell Genet. 2,* 155-64.

Wullems, G. J.; van der Horst, J.; and Bootsma, D. (1976b) Transfer of the human X-chromosome to human-Chinese hamster cell hybrids via isolated HeLa metaphase chromosomes. *Somatic Cell Genet. 2,* 359-71.

Wullems, G. J.; van der Horst, J.; and Bootsma, D. (1977) Transfer of the human genes coding for thymidine kinase and galactokinase to Chinese hamster cells and human-Chinese hamster cell hybrids. *Somatic Cell Genet. 3,* 281-93.

Part V: Plant Tissue Culture

Mutant Selection and Plant Regeneration from Potato Mesophyll Protoplasts

J. F. Shepard

To realize much of the oft-cited potential (Bajaj, 1974; Bhojwani et al., 1977; Kleinhofs and Behki, 1977) of protoplast systems for plant improvement, a number of criteria must first be satisfied. Efficient protoplast isolation and regeneration protocols that are applicable to several cultivars, and certainly not limited to one, are required for the crop plant in question. There must be a means for achieving the desired form(s) of heritable variation in regenerates while otherwise preserving the integrity of the original genome. Finally, selective mechanisms are needed for recognizing and retrieving desirable variants from a large population. For no major crop plant species have all these standards been fulfilled,

Note: The work described herein was supported by National Science Foundation Grants PCM 76-11007, and AER 77-12161; partially by U.S.D.A. Cooperative Agreement 12-14-5001-324; and the Kansas Agricultural Experiment Station. The author gratefully acknowledges the important contributions to this work by postdoctoral fellows Elias Shahin and Dennis Bidney, research assistant Ellen Lozo, and Dr. Larry Williams, Division of Biology, Kansas State University. Also recognized are the cooperative efforts of Dr. J. Pavek and Dr. D. Corsini, U.S.D.A.-SEA Aberdeen, Idaho, and Dr. G. Secor, North Dakota State University, Fargo. Kansas Agricultural Experiment Station Contribution No. 79-278-A.

but conceptual advances from diverse systems suggest that such may soon be possible.

Protoplasts have been prepared enzymatically from either fresh or cultured tissues of several plant species and induced to proliferate in culture (for reviews see Galun et al., 1977; Gamborg, 1977; King et al., 1978). With varying degrees of efficiency, complete regeneration has also been achieved from protoplasts of several of these same plants. Seemingly, it should matter little whether protoplasts are isolated directly from plant parts or indirectly from callus cells so long as sufficiently large populations are obtained. Genetically, however, distinction between the two sources can be important (D'Amato, 1975, 1977; Skirvin, 1978). Karyotypic alterations that accompany the culture of plant cells can eventually cause significant problems in plant regeneration, demonstration of genetic modification, and application to crop improvement. For some species (notably tobacco, but there are others as well) mesophyll protoplasts appear to be especially desirable starting material as judged by the high frequency of normal regenerates (Nagata and Takebe, 1971; Melchers, 1974). Most tobacco plants raised from cells or protoplasts of callus origin, in contrast, were aneuploid or polyploid (Sacristán and Melchers, 1969; Takebe et al., 1971). Hence, the possibility of securing a modified plant that is also suitable for direct application or as a breeding line would be more complicated should the tissue used as a protoplast source have undergone extensive karyotypic overhaul.

Although many techniques for selecting variants in protoplast and/or cell culture have been tested (Dulieu, 1972; Maliga, 1976; Widholm, 1977a), only infrequently have they culminated in plants with a stable genetic modification (Cocking, 1976; Scowcroft, 1977). Within the realm of crop improvement, theoretical and practical limitations of "early" selection are magnified. Mass numbers of potentially distinct genotypes can conveniently be screened at the protoplast stage; but, as the development of the heritable unit progresses, so does the list of identifiable traits. Protoplasts isolated from "crown gall" (transformed by *Agrobacterium tumefaciens)* cells, for example, required exogenous auxin for the initiation of division, and walled single cells did not

(Scowcroft et al., 1973). Hence, attempts (Watts et al., 1975) to achieve and select for transformation with protoplasts might have had more opportunity for success had selection been done after protoplasts had reformed their cell walls.

Some studies have documented a close correlation between the sensitivities of protoplasts (Pelcher et al., 1975), callus cells (Gengenbach et al., 1977), and whole plants to a selective agent (viz., the pathotoxin of *Helminthosporium maydis* race T). Using tissue cultures initiated from Texas male sterile cytoplasm maize, Gengenbach and his associates (1977) successfully selected cell lines resistant to race T toxin. Genetic analysis of progeny from regenerated plants established that resistance to the toxins and the fungus was maternally inherited. However, male sterility, which also characterizes Texas cytoplasm genotypes, was not retained in resistant regenerates or their progeny. Galston and his associates (1977) confirmed that protoplasts from two oat varieties responded differently to another host-specific toxin, victorin (liberated by *H. victoriae*), depending upon the relative sensitivity of the plant to the pathogen. Despite these successes, however, there are as many examples where no constancy of reaction (from protoplast to plant) occurred after exposure to a selective agent or chemical. Beier and her associates (1977) inoculated mesophyll protoplasts of 55 cowpea *(Vigna sinensis)* lines deemed immune to cowpea mosiac virus (CPMV). As protoplasts, all of the lines save one fully supported replication of CPMV. This suggests that expression of viral resistance mechanism(s), although possible, is nevertheless less frequent in protoplasts. In tobacco, some cultivars possess an N gene that at temperatures below 28 °C elicits a necrotic hypersensitive reaction in response to infection by tobacco mosaic virus (TMV). Leaf protoplasts from N gene plants, however, failed to necrose either when inoculated with TMV (Otsuki et al., 1972) or when isolated from plants systemically infected (by maintenance at 30 °C) with the virus (our unpublished results).

At the callus level, essentially the same picture has emerged as per protoplasts. Numerous studies have both confirmed and denied a correspondence of reaction between callus and plant tissues to a pathogen (for a review see Ingram, 1977) or other selective agent. In some cases, response was conditioned by the relative stage of callus development and phenotype (for example, whether white or green in color [Zilkah et al., 1977]). Considering the environmentally modulated response of even mature plants to chemicals such as herbicides and biotic agents, the abovementioned inconsistencies seem inevitable, at least until such time as we can better control the phenotype of the population under selection.

A final means of selection involves first the regeneration of a sizable (around 10,000) plant population and then the analysis of regenerates for specified forms of variation. For obvious reasons, this alternative has received less than an enthusiastic endorsement in the literature. But, should the actual frequency of variation far exceed the anticipated, as is indicated for potato (Matern et al., 1978) and sugarcane (Heinz et al., 1977), the approach should not be altogether unwieldy.

The focus of this book is on the betterment of crop plant species through applications of emerging basic concepts and new technologies. One example in the latter category is the process of whole-plant regeneration from mesophyll cell protoplasts. To date, there is but a single major food crop, the potato, for which this sequence has been reproducibly accomplished. Hence, the focus of this article will be on the potato. For a background on the progress and potential of tobacco (the archetype) and other protoplast systems, several comprehensive reviews are available.

The Problem with Potatoes

Genetically, the potato is a complex and diverse group of tuber bearing species and subspecies belonging to the genus *Solanum*. Collectively, over 160 species are known (Hawkes, 1978), but members of *Solanum tuberosum* L. (mainly the subspecies *tubersum*) occupy cultivated acreage worldwide. Virtually all commercial cultivars are extremely heterozygous tetraploids ($2n = 4X = 48$), which may well have originated not from an autotetraploid as is often portrayed (Ramanna and Wagenvoort, 1976) but rather an amphiploid hybrid (Ugent, 1970; Hawkes, 1978). All potato cultivars are propagated asexually through tubers that are modified stems usually developing from stolons at the basal regions of the plant. Although some cultivars are sexually fertile, many are not or are only partially so or belong to defined compatibility groups. Even when true seeds do result from selfings or crosses between cultivars, none of the progeny closely resembles the parent(s). To confound the problems of genetic analysis, hyperheterozygosity is an acknowleged prerequisite for plant vigor, and F_2 progeny, for example, are substantially less thrifty than their parents (Mendoza and Haynes, 1973). For flowering the subspecies *tuberosum* cultivars may either require relatively long photoperiods (around 16 hours) or be less photoperiod sensitive (Driver and Hawkes, 1943). Tuberization is also strongly influenced (although independently from flowering) by photoperiod, especially when in combination with temperature and light intensity. Although most cultivars will eventually set tubers under the longest of days, there is a substantial difference between them in how early in their development tuberization will begin under a relatively long photoperiod. Most will initiate early under short days; the subspecies *andigena* possesses a nearly absolute short-day requirement for the tuberization stimulus (Ewing, 1978). Nutritionally, the potato is a nearly complete natural plant product (qualitatively) (McCay et al., 1975) and is also capable of producing more protein and amino nitrogen per acre than most other food crops (Wade, 1975).

Historically, the potato has contrasted sharply with the cereals and many other important crop plants in its having been quite refractory to specific improvement through conventional breeding techniques. Today, there are no "green revolution" cultivars possessing simultaneous resistance to most major diseases, requisite horticultural traits, and maximal yield potential. Indeed, the potato is commonly omitted from comprehensive treatises on resistance breeding (for example, Nelson, 1973). The problem in this case has not been solely a lack of available disease-resistant germplasm but rather a useful mechanism for its incorporation without a concomitant loss of vigor and/or other essential traits.

It is important to appreciate that susceptibility to disease is a primary limiting factor in potato production internationally.

One need only reflect on the consequences of the late blight epidemic of the mid-nineteenth century in Ireland to appreciate the potential seriousness of potato disease. Improvements in our genetic and chemical resources have made a recurrence of this disease in equivalent magnitude less likely (although not impossible), but annual losses in potato production from plant pathogens are dramatic still. On a worldwide basis, it has been estimated that 22% of the potato crop is lost each year from disease (Cramer, 1967). If one concurs that the 1970 outbreak of southern corn leaf blight (caused by *Helminthosporium maydis* race T), which eliminated 25% of the crop in Corn-Belt states of the United States, was a "disastrous epidemic" (National Academy of Sciences, 1975), the potato faces an equivalent disaster annually. The U.S.D.A. Crop Reporting Board estimated the 1978 American potato crop at 1.15 million acres, with a value approaching $1 billion. A 22% reduction from disease translates into a $220 million loss in this country alone.

Vegetative propagation of the potato through tubers assures the perpetuation of many infections (including all the viruses and mycoplasms) from generation to generation. And, there seems to be no shortage of insect or disease pests in any potential production area, the tropics being the least hospitable. In the United States and Canada, four cultivars make up something more than 70% of all potatoes grown, with the Russet Burbank ranking first at around 39% (Turnquist, 1978). All of these commercial varieties are over 25 years old and possess similar sensitivities to such major diseases as potato leaf roll, spindle tuber, bacterial ring rot, and *Phytophthora* late blight. This is in spite of a concerted effort over the past 40 years to produce cultivars of equivalent or better horticultural quality with a lesser degree of genetic vulnerability.

As an all-around potato in the United States and Canada, the cultivar Russet Burbank has been difficult to displace. Hence, its origin serves to illustrate the serendipitous nature by which potato improvement has often occurred. In 1871, Luther Burbank chanced upon a single seedball on a vine of an Early Rose potato plant. After nearly losing it from a visit ". . . by a bird or some animal passing rapidly through the field," seeds from the fruit gave rise to 26 new varieties. The most generally desirable of these was named the Burbank by J. H. Gregory, and within a few years it became the chief cultivar on the west coast (Grubb and Guilford, 1912). In the early 1900s, a somatic mutant (sport) with darker colored and russeted tubers appeared in one of the Burbank plants. This selection, which has been asexually propagated since (Hardenburg, 1949), became known as the Russet Burbank and soon supplanted its parent. Today, the cultivar is still preferred for baking and other qualities.

While certainly the Russet Burbank must be a better than average potato, it is not free from significant shortcomings. It has rather exacting environmental requirements that limit its geographical distribution, is not immune to any of the most serious disease organisms, is of only marginal value for many types of processing, and is prone to producing malformed tubers in heavy soils or when water availability is irregular. Hence, several new cultivars with Russet Burbank in their parentage have been released. None, however, has remained free from other problems, ones usually proving more limiting to acceptance than those initially corrected. The Norgold Russet, for example, sets netted tubers of more uniform size and shape and is more determinate in growth habit. Despite these benefits, experience has shown the Norgold Russet to be especially vulnerable to the black leg disease caused by the bacterium *Erwinia carotovora,* variety *atroseptica.*

When attempting to "engineer" a potato possessing specified characteristics, the plant breeder faces enormous problems only superficially addressed above. Should resistance to a disease be obtained, a new strain of the pathogen may quickly overcome it. And, in any case, resistance to a single organism, although helpful, still leaves a broad range left to cope with. Tubers consist largely of water and thus are susceptible to a myriad of physiological as well as pathogen-induced disorders when maintained in storage. The relative degree of susceptibility is genetically influenced as is the length of the period a tuber will remain dormant (resist sprouting). In the field, tubers must conform to prescribed standards of shape, size, uniformity, and depth of set, and foliage cannot be irregular in color or morphology lest growers be unable to visually detect and rogue out viral infections. Of course, there can be no compromise in yield. Given the heterozygosity of the potato plant, satisfying each of these criteria while simultaneously introducing genes for a specific improvement is an inefficient, if not impossible, enterprise (Toxopeus, 1959). In the United States, it has often been accomplished from the fortuitous selection of somatic mutants, or sports, for example, Russet Burbank and Red McClure (Asseyeva, 1927; Crane, 1936; Hardenburg, 1949; Howard, 1970). Parenthetically, this has been true for many other vegetatively propagated crop plants as well, including fruit trees, citrus trees (Miller, 1954; Darrow, 1960), and sweet potatoes (Hernandez et al., 1964). Additional approaches, therefore, appear desirable if we are to efficiently enhance selected traits in the potato. With the recent availability of monoploid (n = X = 12) or true haploid potato lines (Breukelen et al., 1975), a promising new avenue for varietal synthesis may be on the horizon. A second alternative may be to utilize the well-recognized variability of cultured cells, especially in light of the tendency of potatoes to undergo potentially valuable somatic change. There are clear examples in which the regeneration of sugarcane (another vegetatively propagated polyploid) cells has resulted in improved cultivars (Nickell and Heinz, 1973), although it is doubtful whether the true implications of this work have been fully appreciated. Should similar results emerge for the potato, prospects for cultured cell/protoplast systems would appear bright indeed.

Previous Potato Regeneration from Single Cells or Protoplasts

Only very recently have techniques emerged whereby plants may be regenerated from single cells of potato, whether of mesophyll protoplast or cultured cell origin. At the outset, it is important to remember that under the general category of potato are included a range of dihaploid (2n = 2X = 24) lines of *S. tuberosum* L. subspecies *tuberosum* and numerous other species of *Solanum* in addition to commercial cultivars. The distinction between cultivars and other forms of potato, although recipient of little attention in the literature, can nonetheless be significant when defining conditions for protoplast culture and plant regeneration. For example, we have been unable to achieve shoot morphogenesis with protoplast-derived calli (p-calli) of several commercial cultivars by methods (cited below) useful for other types of potato.

The initial report of protoplast isolation for potatoes was with tuber tissues (Lorenzini, 1973), but it was not until 2 years later (Upadhya, 1975) that *in vitro* proliferation was achieved. Leaf-derived protoplasts of the cultivar Sieglinde divided with moderate (around 30%) efficiency, and many resultant calli produced roots (although not shoots). Behnke (1976) was the first to report plant regeneration from single cells, using suspension cultures of six dihaploid potato clones.

In 1977, the rate of progress in potato systems began to accelerate. We (Shepard and Totten, 1977) published protocols for the regeneration of morphologically normal plants from mesophyll protoplasts of the commercial cultivar Russet Burbank. Lam (1977) stimulated embryogenesis and then raised plants from single callus cells of the variety Superior. Binding and Nehls (1977) obtained plants from leaf protoplasts of *S. dulcamara* L. Butenko and his associates (1977a) reported some success with mesophyll protoplasts of *S. chacoense* Bitt., and both callus and leaf protoplasts of *S. tuberosum.* In later abstracts, they claimed plant regeneration from protoplasts of a potato cultivar (Butenko et al., 1977b), but subsequently (Butenko and Kuchko, 1978) found the few regenerates to be aberrant in morphology and chromosome number. In 1978, plants were regenerated from leaf protoplasts of dihaploid *S. tuberosum* clones (Binding et al., 1978; Nehls, 1978a), *S. nigrum* (Nehls, 1978b), and a diploid clone resulting from progeny of a *S. phureja* × *S. chacoense* cross (Grun and Chu, 1978). Most recent in this context is evidence in the exciting paper by Melchers and his associates (1978) for the regeneration of a somatic hybrid between mesophyll protoplasts of tomato (*Lycopersicon esculentum* Mill.) and callus-derived protoplasts of a dihaploid potato clone.

Protoplast Isolation, Culture, and Regeneration—Current Methods

To enhance the reproducibility and efficiency of potato mesophyll protoplast systems, we have continually sought to refine existing techniques and, where necessary, synthesize additional ones. Results from these efforts have led to the adoption of media and protocols superior to those previously reported (Shepard and Totten, 1977) with apparent applicability to other cultivars. In the following paragraphs, current methods used in our laboratory for protoplast isolation and regeneration are given in detail. When precisely followed, these sequences have proved extremely reliable.

The same potato virus X-free clone of *S. tuberosum* L. cultivar Russet Burbank (Shepard and Totten, 1977) served as a primary model in the present study. In addition, the cultivars Butte and Targhee and the longstanding U.S.D.A. seedling 41956 were examined in comparative studies. Tubers were planted in trays of lightly moistened, autoclaved sand. When rooted shoots arising therefrom reached a height of around 10 cm, they were excised, planted in vermiculite in 20-cm pots, and fertilized with the mineral salt solution of Murashige and Skoog (1962) containing twofold amounts of $CaCl_2$ and KH_2PO_4. Plants were maintained in environmental chambers at 18 °C under 12-hour photoperiods of 4000-lux, white fluorescent light and 80% relative humidity, and they were kept well moistened with deionized water. When actively growing, plants were given additional aliquots of the fertilizing solution at 2-week intervals. This plant-growth regimen was developed through extensive factorial analysis of environmental and nutritional components as they influenced the propensity of ensuing protoplasts to proliferate in culture. Without such standardization, protoplast preparations displayed little inclination for division whatever the culture medium employed.

For use in protoplast isolation, petioles with well-expanded terminal leaflets (more than 7 cm long) were removed from young plants (less than 50 cm tall) and preconditioned as follows. Leaves were positioned top side down on a distilled water solution with 2 and 1 mg/1, respectively, of napthalene acetic acid (NAA) and 6-benzylaminopurine (BAP), and 1 m*M* each $CaCl_2$ and NH_4NO_3.

The leaf-flotation step was performed in light-proof vessels at room temperature for 48 hours. Thereafter, leaflets were surface sterilized, lightly brushed to facilitate enzyme infusion, and soaked overnight at 4 °C in a dilute solution of mineral salts as before (Shepard and Totten, 1977). The latter procedure served to "stabilize" ensuing protoplast preparations and has provided for higher and more regular survival percentages during the first 48 hours of protoplast culture. Protoplasts were prepared in a Macerozyme-Cellulase mixed-enzyme solution and collected by flotation in Babcock bottles (Shepard and Totten, 1977).

Float purification has been very effective for preparing potato protoplast preparations essentially free from debris and moribund cells (these are sedimented during centrifugation). For flotation to be a consistently efficient method, two practical considerations deserve mention. The relative degree of cell and, hence, leaf expansion is important. Small terminal leaflets (around 4 cm or less) of potato are typically composed of cells that because of their smaller central vacuoles are not as buoyant in sucrose solutions as are their larger counterparts. Secondly, the concentration of sucrose used for flotation is critical, with 0.3*M* optimal for the system described here. Molarities of 0.35 or more have yielded less reproducible results. Protoplasts isolated from plants (at least as we grow them) undergo a dramatic reduction in volume and buoyancy when transferred from a solution of 0.3*M* to 0.35-0.4*M* sucrose. Apparently, increased protoplast density (from reduced vacuolar volume) at higher osmotic values is not fully offset by simultaneous increases in density of the plasmolyticum (sucrose).

Floated protoplasts were withdrawn from the surface of centrifuged preparations with a pasteur pipet, placed in a 0.3*M* sucrose solution containing one-half the concentration of mineral salts of the CL-medium (Table 1), and again centrifuged at 350 *g* for 10 minutes. Protoplasts were then placed in a small volume of the same rinse solution, their numbers per milliter confirmed with a hemacytometer, and then diluted to 4×10^5/ml. Preparations were incubated for at least 1 hour before plating to allow their adjustment to the higher osmotic and mineral salt levels.

For the culture of potato protoplasts, a combination of two

Table 1. Composition of Cell Layer (CL) and Reservoir (R) Media

Constituent	CL-Medium		R-Medium	
Major Salts				
KNO_3	7600	mg/l	1900	mg/l
$CaCl_2 \cdot 2H_2O$	1760		440	
$MgSO_4 \cdot 7H_20$	1480		370	
KH_2PO_4	680		170	
Iron & Minor Elements				
$Na_2 \cdot$ EDTA	18.5		18.5	
$FeSO_4 \cdot 7H_2O$	13.9		13.9	
H_3BO_3	3.1		3.1	
$MNCl_2 \cdot 4H_2O$	9.9		9.9	
$ZnSO_4 \cdot 7H_2O$	4.6		4.6	
KI	0.42		0.42	
$NA_2MoO_4 \cdot 2H_2O$	0.13		0.13	
$CuSO_4 \cdot 5H_2O$	0.013		0.013	
$COSO_4 \cdot 7H_2O$	0.015		0.015	
Organic Addenda				
Thiamine · HCl	0.5		0.5	
Glycine	2.0		2.0	
Nicotinic acid	5.0		5.0	
Pyridoxine · HCl	0.5		0.5	
Folic acid	0.5		0.5	
Biotin	0.05		0.05	
Casein Hydrolysate	50		100	
Osmoticum				
Sucrose	0.200*M*		0.05	
myo-Inositol	0.025		–	
D-Mannitol	0.025		0.25	
Sorbitol	0.025		–	
Xylitol	0.025		–	
Other				
1-Napthaleneacetic acid	1.0	mg/l	1.0	mg/l
6-Benzylaminopurine	0.4		0.4	
Agar*	0.2%[a]		0.4%[b]	
pH	5.6		5.6	

*Weight:volume. Medium components filter sterilized then mixed with molten agar.

[a] Sigma Type VII Agarose.

[b] Difco Purified Agar, washed.

distinct media has proven the most efficient and reproducible. Cell-layer (CL) medium was that into which protoplasts were dispersed and, hence, initially cultured. Its composition (Table 1) was conducive to the onset of protoplast proliferation although not to continued growth (beyond 2 weeks) of protoplast-derived calli (p-calli). The reservoir (R) medium shown in Table 1 was the most efficient of those tested for sustaining p-callus proliferation after it had commenced. To accomplish a slow exchange between CL- and R-media after protoplast plating, 100-mm plastic quadrant (X-plate) petri dishes were employed. Three slits around 5 mm each in length were made at the base of each septum with a hot number-12 scalpel blade. Then, 10 ml of molten R-medium was added to segments I and III of each dish and allowed to gel. In alternating quadrants (II and IV) were dispensed 3 ml of CL-medium containing protoplasts at between 1.0 and 1.5×10^4/ml. A visual presentation of this system may be seen in Fig. 1. Plates were sealed with parafilm and incubated at 24 °C under 400-lux continuous fluorescent light. Protoplast proliferation began after 4 to 5 days, and plating efficiencies commonly exceeded 50% by the tenth day. As illustrated in Fig. 1, this culture system has been effective for inducing the proliferation of mesophyll protoplasts from not only Russet Burbank but the cultivars Butte and Targhee as well. P-calli were also obtained for the 41956 clone, although protoplast-plating efficiencies were more erratic and rarely exceeded 15%.

The composition of the CL-medium in Table 1 differs substantially from any previously reported for potato systems, and, therefore, certain of its features deserve discussion. The concentrations of major mineral salts (exclusive of NH_4^+) is 4 times that prescribed by Murashige and Skoog (1962) and 40 times that of our earlier report (Shepard and Totten, 1977). While protoplast proliferation has been accomplished in relatively dilute media or in incrementally higher major salt concentrations, greatest reproducibility and efficiency for all cultivars tested has been with the high concentrations of the CL-medium. It was also important to note that KNO_3, KH_2PO_4, $CaCl_2 \cdot 2H_2O$, and $MgSO_4 \cdot 7H_2O$ were simultaneously increased. Factorial experiments in which salt concentrations were varied singly or in combinations of two or

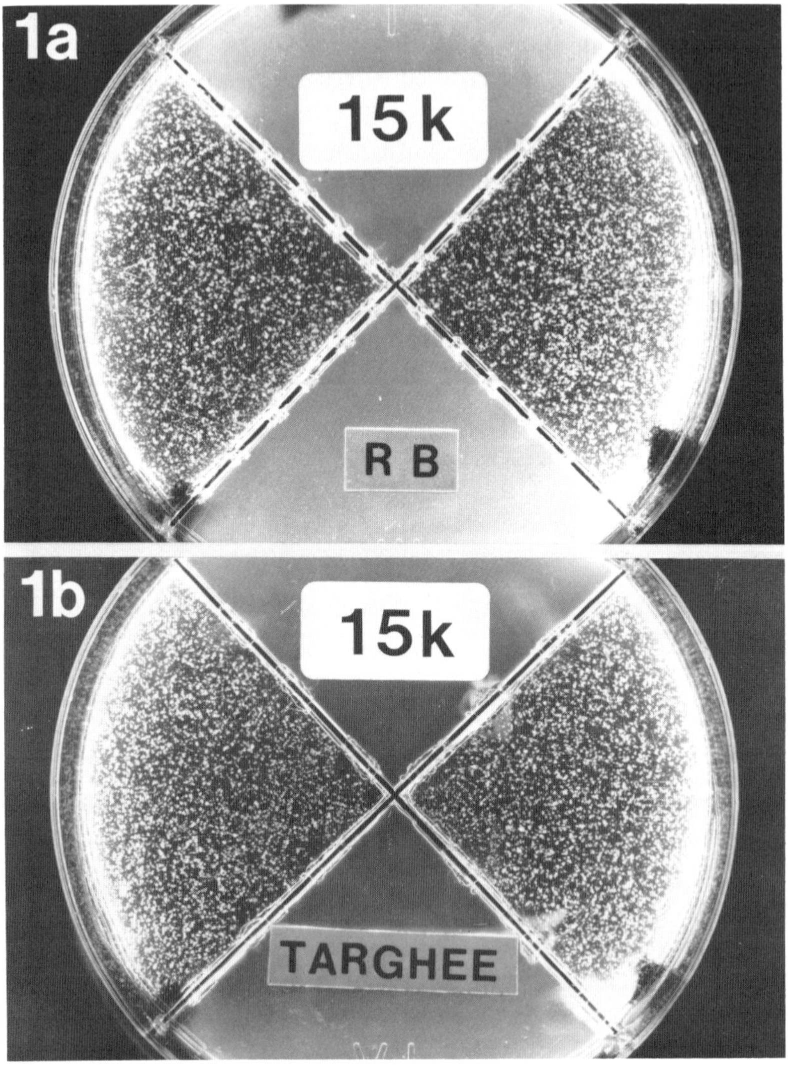

Fig. 1. Photographs of quadrant petri plates 2 weeks after cell-layers were inoculated with mesophyll protoplasts at 15,000/ml (15K). (1a) Plate containing developing calli of Russet Burbank (RB). (1b) Developing calli from Targhee protoplasts.

three always produced lower plating efficiencies than a categorical increase. A similar advantage was observed by blending D-mannitol, myo-inositol, sorbitol, and xylitol with sucrose for the organic osmoticum. Any of these alone or in combinations of two or three were decidedly inferior to the complete mixture. Together (with minor contributions from the remaining medium components), the major mineral salts and organic osmotica produced medium osmotic pressures (determined by freezing-point depression in an Osmette Model A Osmometer) of around 560 mosM, which may be important for maximal results. Formulating the same pressures, however, by increasing organic osmoticum concentrations (whether by one constituent or all five) and lowering mineral salts, or inversely, was detrimental. Thus, if osmotic influence is truly significant in this system, so also must be the balance of components contributing to it. Finally, protoplast and p-callus growth was optimal when very low agar concentrations were used in CL-media. Of several agars tested, Sigma Type VII Agarose (Sigma Chemical Company, St. Louis), after repeated washing with cold distilled water, proved optimal. Key features were its low gelling temperature (25 °C), which facilitated protoplast plating, and its ability to form gels of uniform consistency (albeit very weak) at concentrations of 0.2%. The latter attribute permitted the pipetting of small p-calli during initial transfer.

To regenerate plants from p-calli developing in inoculated culture dishes, a sequence of three additional media was employed: -C, -D, and -E, respectively (Table 2). Both composition of the medium and timing were critical to efficient plant regeneration.

Two weeks after protoplast plating, media in the CL compartments of quadrant petri plates that contained many p-calli, was removed with a small spatula or pipette. Small volumes (around 1 ml) were evenly spread over the surface of culture dishes containing 12 ml of medium-C. Within two weeks, p-calli became green in color and 0.1-1.0 mm in diameter. They were then individually transferred to fresh medium-C for 14 days and then to medium-D for the induction of shoot morphogenesis. Prolonged incubation in protoplast culture (CL) media or C-medium containing more than 0.1 mg/l NAA reduced the capacity of calli to develop shoots on medium-D. Once small (1-2 mm long) shoots had formed

Table 2. Composition of Culture Media C, D, and E

Constituent	Medium-C	Medium-D	Medium-E
NH_4Cl	107 mg/l	267.5 mg/l	267.5 mg/l
KNO_3	1900	1900	1900
$CaCl_2 \cdot 2H_2O$	440	440	440
$MgSO_4 \cdot 7H_2O$	370	370	370
KH_2PO_4	170	170	340
$Na_2 \cdot$ EDTA	18.5	18.5	18.5
$FeSO_4 \cdot 7H_2O$	13.9	13.9	13.9
H_3BO_3	3.1	3.1	3.1
$MnCl_2$·: $4H_2O$	9.9	9.9	9.9
$ZnSO_4 \cdot 7H_2O$	4.6	4.6	4.6
KI	0.42	0.42	0.42
$Na_2MoO_4 \cdot 2H_2O$	0.13	0.13	0.13
$CuSO_4 \cdot 5H_2O$	0.013	0.013	0.013
$COSO_4 \cdot 7H_2O$	0.015	0.015	0.015
myo-Inositol	100 mg/l	100 mg/l	100 mg/l
Thiamine · HCl	0.5	0.5	1.0
Glycine	2	2	2
Nicotinic acid	5	5	5
Pyridoxine · HCl	0.5	0.5	0.5
Folic acid	0.5	0.5	0.5
Biotin	0.05	0.05	0.05
Casein Hydrolysate	100	100	–
Adenine sulfate	40	80	40
1-Naphthaleneacetic acid	0.1	–	0.05
Indole-3-acetic acid	–	0.1	–
6-Benzylaminopurine	0.5	–	–
Zeatin	–	0.5	–
Gibberellic acid[a]	–	–	0.2
D-Mannitol	0.3M	0.2M	0.1M
Sucrose[b]	0.25%	0.25%	1%
MES[c]	5 mM	5 mM	5 mM
Agar[d]	1%	1%	0.5%
pH[e]	5.6	5.6	5.6

[a] Gibberellic acid added after autoclaving.
[b] Weight:volume.
[c] 2-N-Morpholino-ethane sulfonic acid.
[d] Weight:volume, Difco Purified Agar.
[e] pH after autoclaving.

(normally beginning after 4-6 weeks), p-calli were removed from medium-D and pushed to the bottom of a soft layer of medium-E (20 ml in a 100 × 25 mm petri dish). Shoot elongation and root development usually occurred within 3 weeks. Where roots were not initiated by the time shoots had grown to 25 mm in length, the shoot was excised and the cut end immersed in fresh medium-E. Rooting then proceeded within 7 days. Environmental conditions for culture on media-C, -D, and -E were essentially those previously described (Shepard and Totten, 1977). When rooted, plantlets were placed in 3-cm pots containing a 3:1 mixture of vermiculite-sand and moistened with a dilute (0.1 g/l) 20-20-20 soluble fertilizer. Plantlets were maintained at 80% relative humidity in growth chambers under 12-hour photoperiods of 5000-lux fluorescent illumination. When actively growing, plants were transferred to a soil-vermiculite mix.

Forms of Variation

As is true for essentially all other systems where plant populations have been regenerated from single cells, a variable percentage (depending upon the preparation) of protoplast-derived potato plants were categorized as "wild aberrants." These individuals suffered from reduced vigor and exhibited alterations in intensity of green color, number, morphology, and surface area of leaves and formation of aerial tubers, et cetera. Many such deviants could be identified early (sometimes on medium-E) and eliminated. Cursory cytological examination suggests that most, if not all, of these variants possessed other than the normal chromosome complement. For purposes of crop improvement, however, alterations other than the above are the ones of primary interest. Hence, the remainder of this section will be devoted to some examples of variation we have observed among regenerates of Russet Burbank protoplasts that are distinct from the wild-aberrant category. At the same time, this partial list must be considered preliminary until several more tuber generations have been accomplished and confirming data have been collected.

We are at present evaluating two populations of plants that

originated from mesophyll protoplasts. The first began in 1977 as 1700 clones planted at a Montana location. Following field evaluations and harvest, 396 of the original clones were selected, and in 1978 plants were grown at two sites in Idaho by cooperators J. Pavek and D. Corsini, U.S.D.A., Aberdeen, Idaho. Following preliminary evaluations of horticultural performance including yield, tuber size and conformation, maturity date, et cetera, 69 clones were selected as having the greatest potential. Samples of each were planted during the winter of 1978-79 at Homestead, Florida, by G. Secor and D. Johansen, North Dakota State University, Fargo; the remainder are to be extensively evaluated in replicated trials during the summer of 1979. Thus, for this population of 69 clones, we have observed horticultural characteristics over three planting seasons in as many locations. For 396 clones, we have accumulated 2 years of data. As a result, we have been able to identify with confidence several forms of variation within the regenerated population that to this point appear stable. Examples include the following.

Plant Geometry

The normal Russet Burbank cultivar is relatively indeterminate in growth habit, and stems of a typical plant become quite long with substantial internodal distances. Among regenerated clones, however, there were numerous examples in which plants tended toward compactness and possessed shorter internodal lengths. Some clones were quite determinate (Fig. 2), while others ranged between this extreme and the wild type. In still other examples, indeterminacy was maintained but shorter internodes led to the development of a fuller foliar canopy (Fig. 3). Thorough quantitation of these traits for the 69 clones will be completed during the summer of 1979.

Tubers

Under suboptimal conditions of soil texture or moisture supply, Russet Burbank plants can produce a high frequency of malformed, or off-type, tubers. Field observations in 1978 at one

2a
134

2b

Fig. 3. Contrasting foliage of clone M-773 and the parent Russet Burbank (RB). Photograph was taken near the end of the growing season, and differences were present in intensity of green color as well as canopy development.

location revealed that, for a few protoclones (for example, clone M-1), this flaw was manifest to the extent that over 90% of the tubers became malformed. The tendency among regenerates, however, appeared to be toward greater uniformity in tuber conformation, with some clones producing no malformed tubers (Fig. 4). Over 50% of the tubers from parent plants of Russet Burbank were off-type in the same plot. A few clones also differed from the parent in basic tuber shape, and one produced tubers of a considerably greater length-to-width ratio than the norm. Numbers and pounds of tubers per plant (in 10 hill units) varied considerably with the individual clone. Three clones produced twice the number of tubers (and none were malformed) and 80% more weight than the parent. The yield data, however, must await further replication before their significance can be firmly established.

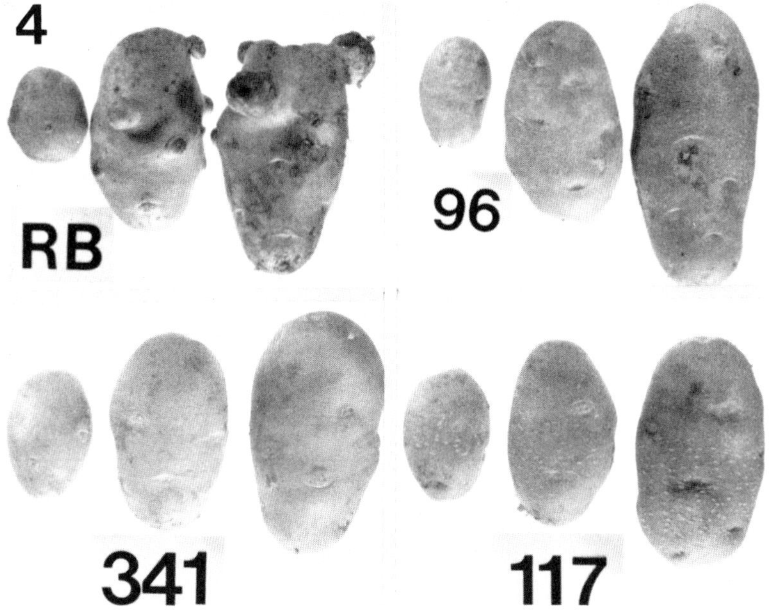

Fig. 4. Representative tubers from Russet Burbank (RB) plants and three protoplast-derived clones, from 1978 test plots in Idaho. Differences are evident between the protoclones in lenticle development (M-117), depth of "eyes" (M-341), and basic tuber shape (M-96), as well as lack of malformation.

Earliness and Depth of Tuber Set

Under a given set of environmental conditions potato cultivars differ from one another in the earliness with which they initiate the tuber-formation process. Those that set tubers relatively late often require a longer growing season to achieve maximal yields and maturity. The soil depth at which tubers are established is similarly under genetic influence and related to the length of the basal stolons. A tuber set too shallow is undesirable because of exposure to sunlight, et cetera, and problems arise during harvest if set is too deep. A detailed analysis of these two traits has not been performed for all of the clones grown in 1978. However, monthly observations throughout the growing season suggested clonal differences at both Idaho locations in how soon after emergence tubers were first detected. Examples were found ranging from somewhat earlier than the parent to considerably later. At harvest, a similar range of variation was suggested in depth of tuber set. Some clones (all units therein) produced tubers that were deeper in the soil than for the normal Russet Burbank, while others tended to form tubers nearer the soil surface.

Seedballs

Russet Burbank is self-sterile under natural conditions and only occasionally will seedless fruits develop to maturity. Flowers and/or small seedballs usually abscise very early in their development. Ten of the protoplast-derived clones consistently produced larger diameter (by as much as 50%) and more frequent (although sterile) berries at all test locations. Some plants became heavily loaded with seedballs that remained attached to the vines long after all on the wild-type had detached. For these variants, the tendency to abscise was diminished, but we have not studied whether or not there was a simultaneously greater stimulus for berry initiation. Because retention of fertilized fruit is a valuable trait in potato breeding, certain clones may be more amenable to improvement efforts as a female parent than the wild-type cultivar.

Resistance to Early Blight Disease

Recently, 500 protoplast-derived clones regenerated in our laboratory were assayed for relative sensitivity to presumptive metabolites of the fungus *Alternaria solani* (Ell. and Marten) Jones and Grout, which causes the early blight disease of potato. Matern and Strobel prepared crude-culture filtrates of the pathogen and found that many regenerates differed from the parent in response to applications of "semipurified toxin" and subsequently to the fungus itself (Matern et al., 1978). Clones that ranged from extremely susceptible (for example, M-419) to resistant (for example, M-742) were identified. Variation in pathogen sensitivity was expressed in progeny of the first and then a second-tuber generation, suggesting that the trait is at least clonally propagated.

Variegation

In the 1977 field plots, protoplast-derived clone (protoclone) M-774 displayed variegated foliage in plants from 2 of the 10 hill units planted. Tubers were collected from variegating stems, and plants have been propagated from them since. Individual "eyes" from a single tuber have produced all-green, all-yellow (bleaching to white), and variegating stems in variable proportions, depending on the tuber. An example of M-774 variegation is shown in Fig. 5. The plant is otherwise typical of Russet Burbank in form and shape. Because of sterility problems in the cultivar, maternal inheritance has not been examined, and, hence, it remains to be established whether the clone is a strict plastosome mutant, a chimera, both, or none of these. Observation of stomatal guard cells over bleached tissues, however, has not revealed the existence of fully developed green plastids.

Clone M-774 was of interest to us as a possible morphological marker for additional protoplast studies, including fusion. Accordingly, we cultured protoplasts from both green and nongreen tissues of this clone. Upon doing so, Dr. Larry Williams observed that, under defined population densities, protoplasts from white tissue areas did not require an exogenous supply of casamino acids for growth. Protoplasts from green tissues of M-774 or normal Russet Burbank, in contrast, displayed an absolute casamino acid requirement at 10,000 protoplasts per milliliter or less (see Fig. 6). Hence, nongreen tissues of the M-774 clone provide both morphological and nutritional markers of potential use in other types of protoplast experiments.

Late Blight

At the time this article was written, a second population of 2000 protoclones was undergoing tuber set in the greenhouse preparatory to field increase and evaluation. None of these clones diverge greatly from the normal growth habit of the Russet Bur-

Fig. 5. Variegation in protoclone M-774.
(Courtesy Dr. Larry Williams.)

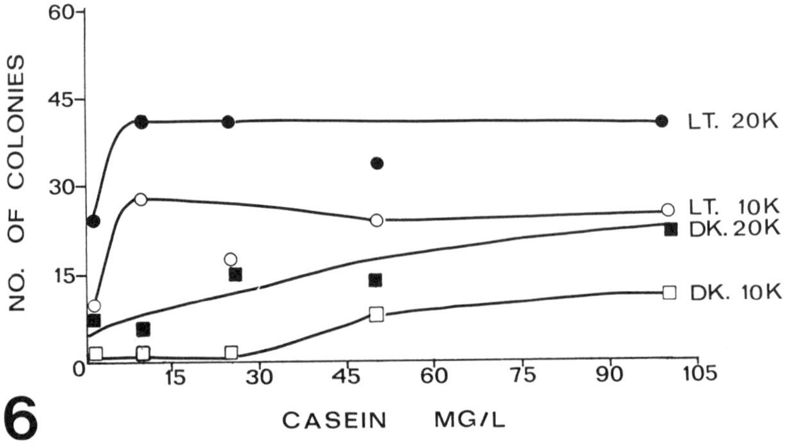

6

Fig. 6. A graph relating protoplast proliferation from yellow (LT.) and green (DK.) leaf tissues with exogenous supply of casamino acids at 10,000 (10K) and 20,000 (20K) protoplasts/ml. Efficiency of proliferation is expressed as number of colonies produced per cubic centimeter. Protoplasts from yellow tissue areas displayed a minor requirement (if any) when their numbers were precisely controlled.

bank cultivar, but smaller differences are nonetheless apparent in some. Postdoctoral fellow Elias Shahin analyzed each of these plants for possible resistance to the late blight fungus *Phytophthora infestans.* In preliminary experiments, several clones that appeared extremely resistant to either race 0 or race 1, 2, 3, or 4 (supplied by H. D. Thurston, Cornell University, and J. Kuć, University of Kentucky, respectively) of the fungus were identified. Only a few scattered, small necrotic flecks appeared on inoculated leaves in contrast to the systemic invasion and death characteristic for the parent clone (Fig. 7). Other protoclones permitted sporulation of *P. infestans* on inoculated leaves, but movement out of the petiole and into the stem was not observed. Rather, infected leaves abscised, and the plantlet continued to grow and develop. At the other extreme, a few clones were systemically invaded and killed more rapidly than was the parent. Neither a comparable range of variation nor examples of resistance were observed in large populations of vegetative cuttings from normal Russet Burbank plants. These experiments have been reproduced at least three times with vegetative cuttings of first-generation clones. Tubers have now been collected from those exhibiting moderate and extreme resistance to ascertain whether corresponding reactions will be expressed in tuber tissues and foliage of the succeeding generation.

A genetic basis for the forms of variation we have observed in potato plant populations regenerated from protoplasts remains to be elucidated. An obvious first thought is that aneuploidy and/or polyploidy may account for most of the variants, especially since these aberrations are frequent in plants raised from callus culture. Indeed, examination of metaphase chromosomes from several variant plants revealed that, for the wild aberrants, chromosome numbers in root-tip cells differed significantly from the 48 typical of the Russet Burbank. This, however, was not true for three clones with the most promise for higher yield potential and improved tuber type (viz., M-96, M-117, and M-341). If a quantitative or qualitative departure from the normal karyotype did occur, it was not sufficient for us to detect. And, from previous work (Kessel and Rowe, 1974), even where trisomy or other minor forms of aneuploidy take place in tetraploid potato, recognition on the basis of phenotype is often not possible. Other genetic

Fig. 7. Differential responses between plants of Russet Burbank (RB) and a resistant (RES) protoclone (K-2784) to foliar inoculation with race 0 of *Phytophthora infestans.* (7a) Seven days after plants were inoculated with zoospores. (7b) Leaves left to right from inoculated RB and K-2784, respectively, 4 days after inoculation. Note extensive disease development and sporulation on RB versus local necrotic reactions on K-2784.

mechanisms, therefore, deserve consideration (see Sunderland, 1977), including the induction of mutation by flavones (Hardigree and Epler, 1978) or other substances that are liberated by cells in culture (Hosel et al., 1977). Because potatoes are highly heterozygous and simplex (Aaaa) for most recognized characters (Howard, 1970), expression of aaa should, or at least could, result from mutation at the A locus. Tetrapolidy per se, therefore, would not preclude manifestation of mutations although such might be indirect, that is, allowing expression of recessive alleles. We are now attempting to regenerate plant populations from mesophyll protoplasts of several desirable protoclones to determine whether or not the same range of variation (for example, resistance to *P. infestans*) will occur within a second protoplast (P_2) generation. Results from this may provide additional insight and avenues of approach.

Prospects for Protoplast and/or Callus Selection

The foregoing discussion suggests that frequent examples of potentially valuable variation exist within potato plant populations raised from mesophyll protoplasts. Consequently, it is reasonable to expect that some variants could be selected early as either protoplasts or small calli and then be regenerated into plants possessing a predicted modification. At the present time, there are no published accounts of this for potatoes, but, from systems becoming available, ones should be forthcoming.

Phytotoxic compounds (toxins) are liberated by some plant pathogenic organisms and in a few instances have been directly linked to the elicitation of disease (reviewed by Scheffer, 1976). As selective agents for securing resistant cells and ultimately resistant plants, such compounds are of special interest. In potato, toxin involvement has been proposed for a number of diseases. Strobel and his coworkers reported a phytotoxic component from culture filtrates of *Alternaria solani* (Matern et al., 1978) and the bacterial ringrot agent *Corynebacterium sepedonicum* (Strobel, 1967). Phenylacetic acid and related compounds accumulated to phytotoxic levels in cultures of the fungus *Rhizoctonia solani,* alternaric acid was liberated by some isolates of *A. solani* (Frank and Francis, 1976), and cell-wall glucan preparations from *P. infestans* specifically agglutinated and killed potato leaf protoplasts (Peters et al., 1978). Application of any of these findings to protoplast selection, however, remains uncertain. For example, we found that toxin preparations from *C. sepedonicum* (supplied by G. Strobel) up to 2 mg/ml were not lethal to mesophyll protoplasts of Russet Burbank after 7 days incubation (unpublished results). We have also encountered difficulty in attempts to apply the reported (Matern et al., 1978) host-specific toxin complex from *A. solani* to protoplast and callus selection. Protoplasts and p-calli from susceptible (M-419), resistant (M-742), and Russet Burbank plants were equally sensitive when we exposed them to various dilutions of "semipurified" (relatively crude) toxin prepared in the manner developed in Strobel's laboratory. Further purification resulted in too little *specific* activity for evaluation in selective systems. More effective methods for preparation of highly purified toxin(s) from *A. solani* will be necessary, therefore, before its reported host specificity can be applied to cultured cell systems.

Other work on protoplast- and cell-level selection in potato is more encouraging. M. Behnke (personal communication) isolated components from cultures of *P. infestans* and, from resistant callus cell lines, regenerated plants tolerant to the phototoxic preparation. It is not known as yet whether the plants are also resistant to the pathogen, but, if they are, the method offers interesting selective prospects. Any resistance so obtained, however, would seem apart from specific forms presently recognized. Late blight resistance in most commercial cultivars is a simply inherited trait and interactions between resistance (R) genes and races of *P. infestans* are quite specific. Plants with, say, an R_1 gene invasion by $Race_1$ of the fungus but are nonetheless susceptible to most other strains of the fungus. Resistance is characterized by the rapid development of a necrotic hypersensitive reaction on inoculated tissue precluding further growth of the pathogen. Hence, a common feature of resistance is the rapid death of affected cells, and we have observed the same pattern in protoclones from Russet Burbank that ward off races 0 and 1, 2, 3, 4. Attempting to select resistance at the cell level, as in Behnke's work, thus would appear paradoxical. Cell lines expressing traditional resistance mechanisms could be the first eliminated, leading to a negative selection for susceptibility. In previous work (see Ingram, 1977), hypersensitive reactions did occur when callus tissues were inoculated with *P. infestans* but not with sufficient regularity for in-depth study (J. Kuć, personal communication). Alternatively, should hypersensitivity be a tissue-mediated response, as apparently is true, for the TMV system described above, death of single, susceptible (and not resistant) cells would occur from other causes. In any case, the experimental system offers intriguing possibilities not only for selecting disease resistance but also for better defining host-pathogen interactions.

Outside the realm of phytotoxins, Carlson and Widholm (1978) selected cell lines from suspension cultures of Merrimac potato that were resistant to growth inhibition by D,L-5-methyltryptophan. One such line accumulated free tryptophan to a level 48 times greater than the norm. No plants were regenerated in these studies, and it is speculative through which stages in development tryptophan accumulation would continue, but the results do suggest potential approaches toward increasing amino-nitrogen pools in potato.

Finally, it would be useful indeed to be able to perform cell-level selection for a broad range of traits that at the plant level would be manifest horticulturally as yield enhancement, heat tolerance, earlier or later tuber set, et cetera. Many such traits, however, are complex in their inheritance; and, given the tetraploid nature of the potato, prospects of success would logically appear dim. Even so, we have observed such changes in plants of protoplast ancestry verifying that they both occur and are expressed. The problem then would seem to be more one of developing early selection methods to enrich regenerate populations. Any suggested approaches to this end remain strictly hypothetical, but selection for variation in phytohormone requirements may offer some interesting possibilities. Cytokinins, for example, have been correlated with phases of tuber formation in the potato (Palmer and Smith, 1969; Forsline and Langille, 1975), and certainly they have pleiotrophic effects throughout the life of the plant. Although presumably not as closely linked with tuber formation, auxins too are primary regulators of plant growth. Thus, selection of variant protoplast or cell lines with altered requirements for cytokinins and/or auxins followed by their regeneration into plants may yield promising results. Widholm (1977b), for example, secured auxin-independent lines of potato and carrot, and the causes of the prototrophy were variable. Had plants been regenerated, it is possible that differences in phenotype may also have occurred.

Conclusions

Modern techniques employed by plant breeders have led to spectacular gains in crop productivity (especially cereals) worldwide. Closer inspection, however, reveals that, despite these well-publicized successes, progress with many other vital crops has languished. Average yields per hectare of soybeans, for example, have increased only slightly during the past two decades. We continue to rely on potato, sweet potato, fruit tree, grape, citrus, and banana (to name a few) cultivars developed 50 or more years ago. Even within major cereal crops, individual problems persist, and new ones continue to emerge as production practices intensify. New technologies that expand the capability of those teams engaged in crop improvement or increase the speed with which specified modifications are achieved will assume ever greater importance as energy supplies dwindle and the human population increases.

In the preceding paragraphs, I have attempted to outline the status of the potato mesophyll protoplast/plant regeneration experimental system as it pertains to the improvement of this significant crop plant. The results are preliminary, and it will require an additional 2 or more years before we can truly assess the horticultural worth of regenerated material. Biologically and genetically, however, certain misgivings about the utility of such systems appear satisfied. Probably the most significant of these is that tetraploidy per se does not preclude the expression of variation. Since many crop plants are polyploids (wheat, for example, is a hexaploid), this is encouraging, at least for those species known to be highly heterozygous. Moreover, the high frequency of variation in potato regenerates for most of the traits analyzed assures the opportunity of deriving something useful whether for experimental or field purposes. Consider that, in a normal potato breeding program (R. Johansen, North Dakota State University, personal communication), around 60,000 seedlings are raised each year in hopes of discovering one of potential value, and any of its attributes would have appeared at random. Finding desirable variants at a frequency of one or more per hundred regenerates is a considerable improvement. Whether or not similar frequencies will occur with other crop plants remains to be seen. Meanwhile, a concerted effort to achieve protoplast regeneration for cereals, legumes, et cetera, is urgently needed before we can make an assessment.

References

Asseyeva, T. (1927) Bud mutations in the potato and their chimerical nature. *J. Genet. 19,* 1-25.

Bajaj, Y. P. S. (1974) Potentials of protoplast culture work in agriculture. *Euphytica 23,* 633-49.

Behnke, M. (1976) Kulturen isolierter Zellen von einigen dihaploiden *Solanum tuberosum* -Klonen und ihre Regeneration. *Z. Pflanzenphysiol. 78,* 177-81.

Beier, H.; Siler, D. J.; Russell, M. L.; and Bruening, G. (1977) Survey of susceptibility to cowpea mosaic virus among protoplasts and intact plants from Vigna sinensis lines. *Phytopathology 67,* 917-21.

Bhojwani, S. S.; Evans, P. K.; and Cocking, E. C. (1977) Protoplast technology in relation to crop plants: Progress and problems. *Euphytica 26,* 343-60.

Binding, H., and Nehls, R. (1977) Regeneration of isolated protoplasts to plants in *Solanum dulcamara* L. *Z. Pflanzenphysiol. 85,* 279-80.

Binding, H.; Nehls, R.; Schieder, O.; Sopory, S. K.; and Wenzel, G. (1978) Regeneration of mesophyll protoplasts isolated from dihaploid clones of *Solanum tuberosum. Physiol. Plant. 43,* 52-54.

Breukelen, E. W. M. van; Ramanna, M. S., and Hermsen, J. G. Th. (1975) Monohaploids (n=x=12) from autotetraploid *Solanum tuberosum* (2n=4x=48) through two successive cycles of female parthenogenesis. *Euphytica* 24, 567-674.

Butenko, R. G., and Kuchko, A. A. (1978) Cultivation and plant regeneration studies on isolated protoplasts from two potato species. *Abstr. Fourth Intl. Congress of Plant Tissue and Cell Culture,* Aug. 20-25, 1978, p. 62. Calgary, Alberta, Canada.

Butenko, R. G.; Kuchko, A. A.; Vitenko, V. A.; and Avetisov, V. A. (1977a) Production and cultivation of isolated protoplasts from the mesophyll of leaves of *Solanum tuberosum* L. and *Solanum chacoense. Sov. Plant Physiol. 24,* 660-65.

Butenko, R. G.; Kuchko, A. A.; and Vitenko, V. A. (1977b) Isolation, culture and plant regeneration studies on protoplasts from two potato species. *Abstr. Intl. Conf. on Regulation of Development Processes in Plants,* July 4-9, 1977, p. 111. Halle, German Democratic Republic.

Carlson, J. E., and Widholm, J. M. (1978) Separation of two forms of anthranilate synthetase from 5-methyltryptophan-susceptible and -resistant cultured *Solanum tuberosum* cells. *Physiol. Plant. 44,* 251-55.

Cocking, E. C. (1976) A new procedure for the selection of somatic hybrids in plants. In *Cell Genetics in Higher Plants* (D. Dudits, G. L. Farkas, and P. Maliga, eds.), pp. 141-48. Akadémiai Kiadó, Budapest.

Cramer, H. H. (1967) *Plant Protection and World Food Crop Production.* Farben Fabriken Bayer, Ag. Leverkusen.

Crane, M. G. (1936) Note on a periclinal chimera in the potato. *J. Genet. 32,* 72-77.

D'Amato, F. (1975) The problem of genetic stability in plant tissue and cell cultures. In *Crop Genetic Resources for Today and Tomorrow* (O. H. Frankel and J. G. Hawkes, eds.), pp. 333-48. Cambridge University Press, Cambridge.

D'Amato, F. (1977) Cytogenetics of differentiation in tissue and cell cultures. In *Plant Cell, Tissue, and Organ Culture* (J. Reinert and Y. P. S. Bajaj, eds.), pp. 343-57. Springer-Verlag, New York.

Darrow, G. M. (1960) Nature of plant sports. *Amer. Hort. Mag. 39,* 123-73.

Driver, C. M., and Hawkes, J. G. (1943) *Photoperiodism in the Potato.* Imperial Bureau Plant Breeding and Genetics, School of Agriculture, Cambridge.

Dulieu, H. (1972) The combination of cell and tissue culture with mutagenesis for the isolation of morphological or developmental mutants. *Phytomorphology 22,* 283-96.

Ewing, E. E. (1978) Critical photoperiods for tuberization: A screening technique with potato cuttings. *Amer. Potato J. 55,* 43-53.

Forsline, P. L., and Langille, A. R. (1975) Endogenous cytokinins in *Solanum tuberosum* as influenced by photoperiod and temperature. *Physiol. Plant. 34,* 75-77.

Frank, J. A., and Francis, S. K. (1976) The effect of a *Rhizoctonia solani* phytotoxin on potatoes. *Can. J. Bot. 54,* 2536-40.

Galston, A. W.; Adams, W. Jr.; Brenneman, F., Fuchs, Y., Rancillac, M.; Reid, R. K.; Sawhney, R. K.; and Staskawicz, B. (1977) Opportunities and obstacles in the culture of cereal protoplasts and calluses. In *Molecular Genetic Modification of Eukaryotes* (I. Rubenstein, R. L. Phillips, C. E. Green, and R. Desnick, eds.), pp. 13-42. Academic Press, New York.

Galun, E.; Aviv, D.; Raveh, D.; Vardi, A.; and Zelcher, A. (1977) Protoplasts in studies of cell genetics and morphogenesis. In *Plant Tissue Culture and Its Bio-Technological Application* (W. Barz, E. Reinhard, and M. H. Zenk, eds.), pp. 302-22. Springer-Verlag, New York.

Gamborg, O. L. (1977) Somatic cell hybridization by protoplast fusion and morphogenesis. In *Plant Tissue Culture and Its Bio-Technological Application* (W. Barz, E. Reinhard, and M. H. Zenk, eds.), pp. 287-301. Springer-Verlag, New York.

Gengenbach, B. G.; Green, C. E.; and Donovan, C. M. (1977) Inheritance of selected pathotoxin resistance in maize plants regenerated from cell cultures. *Proc. Nat. Acad. Sci., U.S.A. 74,* 5113-17.

Grubb, E. H., and Guilford, W. S. (1912) *The Potato.* Doubleday, Page, and Company, Garden City, New York.

Grun, P., and Chu, L-J. (1978) Development of plants from protoplasts of *Solanum* (Solanaceae). *Amer. J. Bot. 65,* 538-43.

Hardenburg, E. V. (1949) *Potato Production.* Comstock Publishing, Ithaca, New York.

Hardigree, A. A., and Epler, J. L. (1978) Comparative mutagenesis of plant flavonoids in microbial systems. *Mutation Res. 58,* 231-39.

Hawkes, J. G. (1978) Biosystematics of the potato. In *The Potato Crop: The Scientific Basis for Improvement* (P. M. Harris, ed.), pp. 15-69. Chapman & Hall, London.

Heinz, D. J.; Krishnamurthi, M.; Nickell, L. G.; and Maretzki, A. (1977) Cell, tissue and organ culture in sugarcane improvement. In *Plant Cell, Tissue, and Organ Culture* (J. Reinert and Bajaj, Y. P. S., eds.), pp. 3-17. Springer-Verlag, New York.

Hernandez, T. P.; Hernandez, T. P.; and Miller, J. C. (1964) Frequency of somatic mutation in several sweet potato varieties. *Proc. Amer. Soc. Hort. Sci. 85,* 430-34.

Hösel, W.; Burmeister, G.; Kreysing, P.; and Surholt, E. (1977) Enzymological aspects of flavonoid catabolism in plant cell cultures. In *Plant Tissue Culture and Its Bio-Technological Application* (W. Barz, E. Reinhard, and M. H. Zenk, eds.), pp. 172-77. Springer-Verlag, New York.

Howard, H. W. (1970) *Genetics of the Potato.* Springer-Verlag, New York.

Ingram, D. S. (1977) Applications in plant pathology. In *Plant Tissue and Cell Culture,* 2nd ed. (H. E. Street, ed.), pp. 463-500. University of California Press, Berkeley.

Kessel, R., and Rowe, P. R. (1974) Interspecific aneuploids in the genus *Solanum. Can. J. Genet. Cytol. 16,* 515-28.

King, P. J.; Potrykus, I.; and Thomas, E. (1978) *In vitro* genetics of cereals: problems and perspectives. *Physiol. Vég. 16,* 381-99.

Kleinhofs, A., and Behki, R. (1977) Prospects for plant genome modification by nonconventional methods. *Ann. Rev. Genet. 11,* 79-101.

Lam, S. (1977) Regeneration of plantlets from single cells in potatoes. *Amer. Potato J. 54,* 575-80.

Lorenzini, M. (1973) Obtention de protoplastes de tubercule de pomme de terre. *C.R. Acad. Sci. Paris 276,* 1839-42.

McCay, C. M.; McCay, J. B.; and Smith, O. (1975) The nutritive value of potatoes. In *Potato Processing* (W. F. Talburg and O. Smith, eds.), pp. 235-73. Avi Publishing, Westport, Connecticut.

Maliga, P. (1976) Isolation of mutants from cultured plant cells. In *Cell Genetics in Higher Plants* (D. Dudits, G. L. Farkas, and P. Maliga, eds.), pp. 59-77. Hungarian Academy of Sciences, Budapest.

Matern, U.; Strobel, G.; and Shepard, J. (1978) Reaction to phytotoxins in a potato population derived from mesophyll protoplasts. *Proc. Nat. Acad. Sci., U.S.A. 75,* 4935-39.

Melchers, G. (1974) Haploids for breeding by mutation and recombination. In *Polyploids and Induced Mutations in Plant Breeding,* pp. 221-31. IAEA, Vienna.

Melchers, G.; Sacristán, M. D.; and Holder, A. A. (1978) Somatic hybrid plants of potato and tomato regenerated from fused protoplasts. *Carlsberg Res. Commun. 43,* 203-18.

Mendoza, H. A., and Haynes, F. L. (1973) Some aspects of breeding and inbreeding in potatoes. *Amer. Potato J. 50,* 216-22.

Miller, E. V. (1954) The natural origins of some popular varieties of fruit. *Econ. Bot. 8,* 337-48.

Murashige, T., and Skoog, F. (1962) A revised medium for rapid growth and bioassays with tobacco tissue cultures. *Physiol. Plant. 15,* 473-97.

National Academy of Sciences. (1975) Pest control: An assessment of present and alternative technologies. In *Corn/Soybeans Pest Control,* vol. II. National Academy of Sciences, Washington, D.C.

Nehls, R. (1978a) Isolation, culture, and regeneration of protoplasts from Solanaceous species. *Abstr. Fourth Intl. Congress of Plant Tissue and Cell Culture,* Aug. 20-25, 1978, p. 62. Calgary, Alberta, Canada.

Nehls, R. (1978b) Isolation and regeneration of protoplasts from *Solanum nigrum* L. *Plant Sci. Lett. 12,* 183-87.

Nelson, R. R. (1973) *Breeding Plants for Disease Resistance: Concepts and Applications.* Pennsylvania State University Press, University Park, Pennsylvania.

Nickell, L. G., and Heinz, D. J. (1973) Potential of cell and tissue culture techniques as aids in economic plant improvement. In *Genes, Enzymes and Populations* (A. Srb, ed.), pp. 109-28. Plenum Press, New York.

Otsuki, Y.; Shimomura, T.; and Takebe, I. (1972) Tobacco mosaic virus multiplication and expression of the N gene in necrotic responding tobacco varieties. *Virology 50,* 45-50.

Palmer, C. E., and Smith, O. E. (1969) Cytokinins and tuber initiation in the potato *Solanum tuberosum* L. *Nature 221,* 279-80.

Pelcher, L. E.; Kao, K. N.; Gamborg, O. L.; Yoder, O. C.; and Gracen, V. E. (1975) Effects of *Helminthosporium maydis* race T toxin on protoplasts of resistant and susceptible corn *(Zea mays). Can. J. Bot. 53,* 427-31.

Peters, B. M.; Cribb, D. H.; and Stelzig, D. A. (1978) Agglutination of plant protoplasts by fungal cell wall glucans. *Science 201,* 364-65.

Ramanna, M. S., and Wagenvoort, M. (1976) Identification of the trisomic series in diploid *Solanum tuberosum* L., group tuberosum. I: Chromosome identification. *Euphytica 25,* 233-40.

Sacristán, M. D., and Melchers, G. (1969) The caryological analysis of plants regenerated from tumorous and other callus cultures of tobacco. *Mol. Gen. Genet. 105,* 317-33.

Scheffer, R. P. (1976) Host-specific toxins in relation to pathogenesis and disease resistance. In *Physiological Plant Pathology* (R. Heitefuss and P. H. Williams, eds.), pp. 247-69. Springer-Verlag, New York.

Scowcroft, W. R. (1977) Somatic cell genetics and plant improvement. *Advan. Agron. 29,* 39-81.

Scowcroft, W. R.; Davey, M. R.; and Power, J. B. (1973) Crown gall protoplasts—isolation, culture and ultrastructure. *Plant Sci. Lett. 1,* 451-56.

Shepard, J. F., and Totten, R. E. (1977) Mesophyll cell protoplasts of potato: Isolation, proliferation and plant regeneration. *Plant Physiol. 60,* 313-16.

Skirvin, R. M. (1978) Natural and induced variation in tissue culture. *Euphytica 27,* 241-66.

Strobel, G. A. (1967) Purification and properties of a phytotoxic polysaccharide produced by *Corynebacterium sepedomicum. Plant Physiol. 42,* 1433-41.

Sunderland, N. (1977) Nuclear cytology. In *Plant Tissue and Cell Culture,* 2nd ed. (H. E. Street, ed.), pp. 177-205. University of California Press, Berkeley.

Takebe, I.; Labib, G.; and Melchers, G. (1971) Regeneration of whole plants from isolated mesophyll protoplasts of tobacco. *Naturwiss. 58,* 318-20.

Toxopeus, H. J. (1959) Problems involved in breeding for resistance. *Euphytica 8,* 223-31.

Turnquist, O. C. (1978) Production of certified seed potatoes by varieties, 1976-77. *Amer. Potato J. 55,* 274-86.

Ugent, D. (1970) The potato. *Science 170,* 1161-66.

Upadhya, M. D. (1975) Isolation and culture of mesophyll protoplasts of potato *(Solanum tuberosum* L.). *Potato Res. 18,* 438-45.

Wade, N. (1975) International agricultural research. *Science 188,* 585-89.

Watts, J. W.; Cooper, D.; and King, J. M. (1975) Plant protoplasts in transformation studies: Some practical considerations. In *Modification of the Information Content of Plant Cells* (R. Markham, D. R. Davies, D. A. Hopwood, and R. W. Horne, eds.), pp. 119-31. Elsevier, New York.

Widholm, J. M. (1977a) Selection and characterization of biochemical mutants. In *Plant Tissue Culture and Its Bio-Technological Application* (W. Barz, E. Reinhard, and M. H. Zenk, eds.), pp. 112-22. Springer-Verlag, New York.

Widholm, J. M. (1977b) Relation between auxin autotrophy and tryptophan accumulation in cultured plant cells. *Planta 134,* 103-8.

Zilkah, S.; Bocoin, P. F.; and Gressel, J. (1977) Cell cultures vs. whole plants for measuring phytotoxicity. II: Correlations between phytotoxicity in seedlings and calli. *Plant Cell Physiol.* 18, 657-70.

Epigenetic Changes in Tobacco Cell Culture: Studies of Cytokinin Habituation

F. Meins, Jr., and J. Lutz

Plant tissues in culture commonly undergo variation. This involves changes in such diverse phenotypes as chromosomal constitution (D'Amato, 1977), requirement for specific growth factors (Gautheret, 1955), capacity for organogenesis (Skirvin, 1978), production of secondary products (Butcher, 1977), and resistance to inhibitors of amino acid metabolism (Carlson, 1973; Widholm, 1974) and of protein and nucleic acid synthesis (Lescure, 1973; Maliga et al., 1973a and b; Sung, 1976). Although some of this variation can be attributed to the selection of specific cell types in the original tissue explant (Gautheret, 1966), it is likely that most variation results from alterations in cellular heredity (Meins, 1975).

Variation in culture is of considerable practical importance. On the one hand, when tissue culture is used as a technique for plant propagation, it is essential to minimize those forms of variation that are transmitted to the daughter plants. On the other hand, natural and induced variation in culture may be exploited to

Note: This work was supported by grant number CA 20053 from the National Cancer Institute and grant number PCM 74-18906 from the National Science Foundation.

obtain varieties of economically important plants. Both applications of tissue culture depend on understanding the nature of cellular variation and on learning how to regulate it.

This paper deals with a type of cellular variation known as cytokinin habituation. As background, we will begin by describing how two sources of variation in culture—epigenetic changes and genetic mutations—can be distinguished experimentally. We will then provide evidence that habituation has an epigenetic basis, that expression of the habituated state is maintained by a positive feedback loop, that certain tissue-specific states of habituation are not erased when cells are induced to regenerate plants, and that different totipotent cells in the same plant may differ genetically.

Distinguishing between Genetic Mutations and Epigenetic Changes

In principle, cellular variation can result from genetic mutation, epigenetic change, or a combination of both processes. Genetic mutations involve rare, random alterations in genetic constitution, such as point mutations, deletions, and chromosomal rearrangements. Epigenetic changes are more difficult to define because their underlying mechanism is as yet unknown. These changes, in contrast to classical mutations, involve directed alterations in cellular phenotype that occur at very high rates, are regularly reversible, and leave the heritably altered cell totipotent (Nanney, 1958; Meins, 1972). Epigenetic changes are well documented in unicellular organisms in which direct genetic analysis of the phenotypically altered cell is possible (for example, Beale, 1958). There is also considerable, although indirect, evidence that these changes play an important role in the determination of new cell types that arise in the development of multicellular organisms (Meins and Binns, 1978).

When variant cells in culture remain totipotent, it is possible to detect mutations by breeding plants derived from the progeny of these cells. If the heritable change results from mutation, then the variant phenotype should persist in the regenerated plants or in tissues cultured from these plants and it should be transmitted in a regular fashion when the plants are crossed. There are numerous reports of putative mutants that arise in culture. Only a few, however, have been studied in sufficient detail to establish that the variants have a genetic basis—for example, methionine sulfoximine (Carlson, 1973), streptomycin (Maliga et al., 1973a), and 5-bromodeoxyuridine (Maliga et al., 1973b) resistance in haploid cell lines of *Nicotiana* and helminthosporium-toxin resistance in tissues of maize (Gengenbach et al., 1977).

The Epigenetic Basis for Cytokinin Habituation

Convincing evidence for epigenetic variation in culture comes from studies of cytokinin habituation (Meins and Binns, 1978). Pith parenchyma tissues excised from tobacco plants require both an auxin and a cell-division factor (CDF), usually provided as a cytokinin, for continued growth in culture (Jablonski and Skoog, 1954). These tissues sometimes lose their exogenous requirement for cytokinin or for both auxin and cytokinin after a variable number of passages in culture (Gautheret, 1955; Fox, 1963). Thereafter, the habituated tissues and clones derived from these tissues can be propagated indefinitely without added growth factor (Lutz, 1971; Binns and Meins, 1973). These tissues also produce significant amounts of the growth factor for which they are habituated (Kulescha and Gautheret, 1948; Wood et al., 1969; Dyson and Hall, 1972; Einset and Skoog, 1973). The fact that certain tissues in the intact plant produce the same growth factors (Thimann, 1969; Feldman, 1975) suggests that habituation involves a stable, cellular change in the expression of genetic potentialities of the pith cell that are normally silent in culture.

We have been studying the nature of the heritable change in cytokinin habituation of pith parenchyma tissues from Havana 425 tobacco. Our choice of this experimental system was a fortunate one because Havana 425 pith cells habituate rapidly, remain organogenic for many transfer generations in culture, and are readily cloned from single cells. Epigenetic changes are directed and occur at high rates at the cellular level relative to classical mutations. Pith tissues rapidly habituate when incubated at 35 °C, about 10 °C above the standard culture temperature (Meins, 1974), or, as we will show later, when incubated on a complete medium containing the synthetic cytokinin, kinetin. Cytokinin-habituated cells can support the proliferation of nonhabituated cells in the same tissue by crossfeeding (Tandeau de Marsac and Jouanneau, 1972; Meins and Binns, 1977). Thus expression of the habituated phenotype does not necessarily provide a reliable estimate of the number of habituated cells in the tissue. To measure the rate at which cells habituate, we used an approach based on the method of Luria and Delbrück (1943) for measuring mutation rates in bacterial populations. When pith explants are incubated at 35 °C on a basal, auxin-containing medium without added cytokinin, the cells enlarge considerably and, hence, the tissues become translucent. Some of these explants also form dense, white, hyperplastic nodules that are clearly discernible by transillumination. These nodules, but not the surrounding translucent tissues, are cytokinin habituated. Thus, the appearance of a nodule provides a marker for the habituation event. If this event occurs at random, then the proportion of replicate explants of the same size without nodules provides an estimate of the number of conversions to the habituated phenotype per explant. This value divided by the number of cell generations during the incubation period is the rate of habituation. Although this rate varied widely from experiment to experiment, the average was $>2.8 \times 10^{-3}$ conversions per cell generation (Meins, 1977). The actual rate is undoubtedly much higher since habituated cells arising late in the induction period probably form nodules too small to detect by eye. The important point is that the rate observed, which is a lower limit, is 100 to 1000 times faster than the rates reported for somatic mutation in tobacco (Sand et al., 1960; Carlson, 1974).

Cytokinin-habituated clones regularly form plants when placed on an inductive medium. For example, in our first experiment, we obtained 51 regenerated plants from 13 habituated clones, and, of these plants, 50 were fertile (Binns and Meins, 1973). When pith tissues from the regenerated plants were placed in culture, they always required an exogenous source of cytokinin for growth and, in this regard, were indistinguishable from pith tissues from seed-grown plants. Two important conclusions may be drawn from these results. First, habituated cells have remained totipotent. Second, at some point during the regeneration of plants, habituated clones have reverted to the nonhabituated state.

Because plants were regenerated from clones that had been subcultured several times, it may be argued that nonhabituated cells arise in the tissues by very low rates of back mutation and that it is these cells that eventually form plants. There is strong evidence against this. First, we have subcloned habituated lines and immediately placed the subclones on a bud-inducing medium. In a typical experiment, 33 of 45 subclones formed plants, indicating that plants did not develop from revertant cells arising during prolonged culture. Second, the rate of bud initiation from recently isolated habituated and nonhabituated subclones was roughly the same. Finally, the rate of bud initiation of subclones from the same clone of habituated tissue was $8.2 \pm 5.8 \times 10^{-3}$ initiations per cell generation (± standard error, 7 subclones). This value, which is a minimum estimate, is at least 1000 times faster than would be expected were bud initiation to depend upon back mutation. Apparently, plants either develop directly from habituated cells or from nonhabituated cells that arise in the tissues at very high rates.

The experiments summarized here provide strong evidence that cytokinin habituation of tobacco pith cells has an epigenetic basis. First, induction and reversal of habituation are directed processes, that is, they occur in response to specific stimuli. For example, habituation is induced in cultures by growing them in the presence of kinetin or at 35 °C. Second, both induction and reversal of habituation occur at very high rates, considerably faster than classical mutations. Third, habituated cells remain totipotent. Finally, habituation involves the expression of a genetic potentiality of the pith cell. Cytokinin habituation offers, therefore, one of the very few experimental systems in which epigenetic changes can be induced and reversed at will under precisely defined conditions in cell culture. The remaining sections of this chapter will deal with experiments aimed at learning how the habituated state is initiated and maintained in a population of dividing cells.

A Positive-Feedback Hypothesis for Habituation

Our working hypothesis, described in detail elsewhere (Meins and Binns, 1978), is that the habituated state is maintained by a positive-feedback loop in which cytokinins or related growth regulators either induce their own synthesis or inhibit their own degradation. The basic idea is that cells in fresh explants of pith are poised to produce cell-division factors (CDF) but require CDF above some critical concentration in order to induce production. When an exogenous source of CDF is provided above this critical concentration, the cells shift to the habituated state and remain in this state so long as the self-sustaining feedback loop is not interrupted.

This hypothesis makes specific predictions, which we have begun to verify experimentally. First, it should be possible to induce habituation by treating tissues with CDF. To find out whether this was the case, we incubated cytokinin-requiring pith explants with increasing concentrations of kinetin and then scored the explants for habituated nodules. Kinetin concentrations below roughly 4×10^{-9} M did not induce habituation (Table 1). Above this threshold, the incidence of habituation increased with higher kinetin concentrations. All of the explants were habituated at a

Table 1. Effect of Kinetin Concentration on the Induction of Habituation in Primary Explants of Pith

Kinetin Concentration (M)[a]	Percent of Explants Habituated[b]
0	0
7.5×10^{-10}	0
3.7×10^{-9}	0
1.9×10^{-8}	21
9.3×10^{-8}	100
4.7×10^{-7}	100

[a]A medium containing the sucrose and salt concentrations recommended by Linsmaier and Skoog (1965) supplemented with 2 mg/α-naphthalene acetic acid. Media further supplemented with kinetin at the concentrations indicated.

[b]Explants incubated for 21 days at 25 °C in diffuse light and scored for hyperplastic nodules; 14 replicates in each treatment.

kinetin concentration of around 5×10^{-7} M, which is near optimal for the growth of nonhabituated tissues. Thus, as predicted by our hypothesis, CDF induces habituation when provided above a critical threshold concentration.

A second prediction is that cells depleted of CDF by blocking CDF production momentarily should return to the nonhabituated state. To verify this, we incubated cloned, habituated tissues for several transfers at 16 °C or on a medium containing a reduced concentration of auxin. Under these conditions, habituated tissues require exogenous cytokinin for growth, suggesting that CDF production is blocked (Syōno and Furuya, 1971; Tandeau de Marsac and Jouanneau, 1972; Einset, 1977; Binns and Meins, 1979). We ran parallel experiments in which tissues under the nonpermissive conditions (that is, 16 °C or low-auxin medium) were incubated with optimal concentrations of kinetin and with low concentrations of kinetin that supported growth but that were below the threshold needed to induce habituation. The tissues were then shifted to permissive conditions (that is, 25 °C or high auxin medium) and were assayed for habituation. In agreement with our working hypothesis, tissues on high-kinetin medium remained habituated when assayed under permissive conditions, whereas tissues on low-kinetin medium shifted to the nonhabituated state. Moreover, these nonhabituated tissues, like pith tissues, regained their habituated character when subcultured for a single transfer on high-kinetin medium under permissive conditions (Meins and Binns, 1978). These results provided strong but indirect evidence that the habituated state was maintained by a positive-feedback loop involving CDF.

We are currently trying to test this hypothesis directly by measuring the concentration of CDF in tissues under conditions that either induce or reverse habituation. If our hypothesis is correct, the concentration of CDF should decrease in tissues during reversal of habituation and tissues induced to habituate by treatment with CDF should produce more of this factor than the amount supplied.

Evidence for Two Different Stable Changes in Habituation

According to the positive-feedback hypothesis, nonhabituated cells should always habituate when treated with high concentrations of kinetin. We found that certain recalcitrant cell types did not. Studies of these cells provided evidence that the habituation process involved at least two different stable changes. Pith, stem-cortex, and leaf tissues isolated from the same plant differ in their cytokinin requirement and capacity to habituate (Table 2). As expected, primary explants of pith initially required cytokinin and habituated after a single transfer on kinetin-containing medium. Cortex tissues were initially habituated and their degree of habituation, expressed as relative growth rate on media with and without kinetin (column A/B in Table 2) did not increase with kinetin treatment. Leaf tissue differed from pith in several ways. First, although this tissue, like pith, initially required cytokinin for rapid growth, the cytokinin requirement was not lost after kinetin treatment. Second, leaf tissue grew, although very slowly, when subcultured on basal medium. Finally, the leaf phenotype was extremely stable in culture. Callus of leaf origin or clones derived from this tissue still required cytokinin for rapid growth after more than 10 transfers on kinetin-containing medium.

Table 2. The Cytokinin Requirement and Capacity to Habituate of Pith, Cortex, and Leaf Tissues of Tobacco in Culture

		Growth[c] after One Transfer		
Tissue	Previous Treatment	A –Kinetin	B +Kinetin	A/B
Pith	–Kinetin[a]	2.21 ± 0.50(8)[d]	33.0 ± 4.68(8)	0.067
	+Kinetin[b]	70.9 ± 0.4(9)	17.0 ± 2.37(9)	4.17
Cortex	–Kinetin	36.7 ± 4.87(9)	45.8 ± 15.8(9)	0.80
	+Kinetin	56.4 ± 6.35(9)	40.3 ± 7.19(9)	1.40
Leaf	–Kinetin	3.07 ± 0.75(9)	153.3 ± 25.4(9)	0.020
	+Kinetin	2.22 ± 0.24(9)	168.9 ± 25.1(9)	0.013

[a]A medium containing the sucrose and salt concentrations recommended by Linsmaier and Skoog (1965) supplemented with 2 mg/l α-naphthalene acetic acid as a source of auxin.

[b]The minus-kinetin medium further supplemented with 0.3 mg/l kinetin.

[c]Growth expressed as (final weight – initial weight)/initial weight after 21 days. For culture conditions, see Meins and Binns (1977).

[d]Mean and standard error for of the mean (n) replicates.

These results bear directly on the problem of how the habituated state is maintained. Pith cells are totipotent and, hence, can give rise to leaf cells. This implies that during organogenesis some progeny of the cultured pith cells have shifted to a new state in which they are unable to respond to CDF by habituating. To test this hypothesis, we induced clones of habituated pith tissue to form plants, incubated leaf tissues from these plants on kinetin-containing medium, and then assayed the leaf callus for habituation. Leaf tissues of pith-cell origin did not habituate after kinetin treatment and, in this regard, were indistinguishable from leaf tissues from seed-grown plants (Table 3). We also performed the reciprocal experiment and found that pith tissues of leaf-cell origin exhibit the phenotype typical of pith in culture.

To summarize the results presented so far, tobacco cells can undergo stable, reversible changes in their competence to respond to CDF by habituating as well as in their capacity to express the habituated phenotype. Pith cells are competent to habituate but only do so in culture when triggered by CDF. We have indirect evidence that the stability of this change is affected by some sort of positive-feedback loop involving CDF. Different states of competence, on the other hand, are extremely stable in culture and do not depend on the expression of the habituated state for their maintenance. For example, pith cells remain competent when induced to revert to the nonhabituated state by interrupting the putative CDF-feedback loop. Changes in competence regularly occur when pith or leaf cells are induced to undergo organogenesis. The mechanism for this second type of stable change is not known.

Hints That Certain Epigenetic Changes in Habituation Have a Genetic Basis

Our interpretation of habituation is predicated on two commonly accepted assumptions: first, that tissue-specific phenotypes are somehow erased when cells are induced to regenerate plants; and second, that different totipotent cells from the same plant are genetically equivalent. These assumptions may not be well founded. In this section, we will present evidence that, in some cases,

Table 3. The Cytokinin Requirement of Tissues from Regenerated Plants of Leaf and Pith Cell Origin.

	Growth of Parent tissue[a]				Growth of Tissue from Regenerated Plant		
Parent Tissue	A −Kinetin	B +Kinetin	A/B	Tissue Assayed[c]	C −Kinetin	D +Kinetin	C/D
Pith (Clone 153H)	82 ± 28(2)[b]	46 ± 15(3)	1.8	Leaf	4.0 ± 0.2(3)	84 ± 11(3)	0.048
Leaf (Clone L20)	14.5 ± 6.0(6)	167 ± 58.4(6)	0.087	Pith	46.3 ± 6.2(9)	31.8 ± 19.9(9)	1.46

[a]Growth measurement and media described in Table 2.

[b]Mean and standard error of the mean (n).

[c]Tissues from plants regenerated from the cloned tissue indicated in column 1. The tissues were assayed after at least one transfer on kinetin-containing medium.

tissue-specific states of habituation persist when cells undergo meiosis. This evidence is preliminary and it is based on the analysis of cells isolated from just one tobacco plant.

Cortex and leaf tissues exhibit different states of habituation in culture (see Table 2). To find out whether or not these states are erased when cells are induced to form plants, we compared the cytokinin requirement of leaf and cortex tissues from plants of leaf and cortex cell origin. The first step was to isolate numerous clones of leaf and cortex tissue from the same plant. The clones were then assayed for degree of habituation defined as the ratio (R), growth (-kinetin)/growth (+kinetin), where growth is expressed as the natural logarithm of tissue weight (W) after 21 days of incubation divided by initial tissue weight (W_0), that is, ln (W/W_0) (Meins, 1974). Tissues with an absolute cytokinin requirement exhibit R values in the range of around 0-0.4, whereas habituated tissues exhibit R values greater than around 0.4 (Meins and Binns, 1977). Based on these criteria, 42 of the leaf clones isolated were not habituated and 2 clones were slightly habituated. The average R value for the 44 clones was 0.20 ± 0.02 (± standard error). All of the cortex clones were habituated with an average R value for 50 clones of 1.15 ± 0.04. Thus, the characteristic leaf and cortex phenotypes are inherited by individual cells.

The clones of leaf and cortex origin regularly formed complete, fertile plants when incubated on a standard inductive medium (Binns and Meins, 1973). We assayed leaf and cortex tissues from the regenerated plants for degree of habituation. Table 4 summarizes results obtained with a seed-grown plant and plants regenerated from 10 leaf and 10 cortex clones. If tissue-specific states of habituation are erased during organogenesis, then tissues from the regenerated plants should exhibit the same degree of habituation as the comparable tissues from seed-grown plants. This was the case for three of the tissues: cortex tissues of leaf as well as of cortex origin were habituated and leaf tissues of cortex origin were cytoleinin requiring. In striking contrast, leaf tissue of cortex origin exhibited the habituated phenotype. These results provide direct experimental evidence that plants regenerated from two different totipotent cell types, in this case leaf and cortex, are not necessarily equivalent. Moreover, because only leaf tissues of cortex origin are

Table 4. Degree of Habituation of Tissues from Regenerated Plants of Cortex and Leaf Origin. (Grand average of 10 experiments)

Source of Plants	Tissue Assayed	Degree of Habituation (R)[b]
Seed grown[a]	Cortex	1.09 ± 0.13(9)[c]
	Leaf	0.25 ± 0.03(8)
Cortex clones	Cortex	1.33 ± 0.24(10)
	Leaf	0.80 ± 0.05(10)*
Leaf clones	Cortex	1.16 ± 0.11(10)
	Leaf	0.27 ± 0.03(10)

[a]Data from Table 1 expressed as R values.

[b]Grand averages for tissues from plants regenerated from 10 different clones. For tissues from the seed-grown plant, the average of replicate tissues.

[c]Mean R values and standard error of the mean for (n) replicates.

*The difference in mean R values of leaves from cortex and leaf derived plants was highly significant (t–test of means, $p<.001$).

habituated, it appears that leaf cells "remember" the habituated character of the cortex cells from which they were derived.

The important question that arises is whether or not the two different leaf phenotypes persist in plants propagated sexually. To answer this, we allowed plants regenerated from cortex and leaf cells to flower, self-pollinate, and set seed. We then germinated the seeds and assayed leaf tissues from each mature plant for degree of habituation. The results obtained with 15 progeny from each of 8 plants derived from 4 leaf and 4 cortex clones are summarized in Table 5. The leaf tissues of progeny from leaf-derived plants, in every case, were nonhabituated. The average R values for the progeny were also in the range obtained with leaf tissue from 15 different seed-grown plants. Leaf tissues of progeny from cortex-derived plants, on the other hand, fell into two classes; based on R values, one class was habituated, the other class was not habituated. The number of progeny in each of these two classes was remarkably consistent from clone to clone. Of the 60 progeny scored, 38, or 63%, exhibited the habituated phenotype. These findings showed that the habituated phenotype characteristic of leaves from plants of cortex origin was transmitted to a large proportion of the pro-

Table 5. The Leaf Phenotype of Progeny from Seed-Grown Plants and Plants Regenerated from Cortex and Leaf Cells

		Habituated[a]		Nonhabituated[c]	
Source of Parent Plants		Number of Progeny	R[b]	Number of Progeny	R
Seed grown		0	–	15[d]	0.38 ± 0.02
Leaf clone	12	0	–	12	0.30 ± 0.02
	33	0	–	15	0.28 ± 0.02
	34	0	–	15	0.34 ± 0.02
	35	0	–	15	0.24 ± 0.02
Cortex clone	1	10	0.66 ± 0.03	5	0.23 ± 0.02
	41	8	0.71 ± 0.04	7	0.11 ± 0.06
	44	9	0.73 ± 0.05	6	0.20 ± 0.05
	45	11	0.72 ± 0.03	4	0.20 ± 0.08

[a]Progeny with leaf tissues exhibiting $R>0.4$.

[b]The average obtained with 8 replicate leaf explants for the number of progeny stated.

[c]Progeny with leaf tissues exhibiting $R<0.4$.

[d]Leaf tissues from 15 seed-grown plants chosen at random.

geny from these plants. Moreover, since some of the progeny of the self-cross have wild-type leaves, it appears that the habituated phenotype of leaves is potentially reversible.

Conclusions

Our experiments provide strong evidence that cytokinin habituation of tobacco pith cells is a directed, potentially reversible process that leaves the heritably altered cells totipotent and results in the expression of latent genetic potentialities. Thus, it appears that habituation has as its basis epigenetic rather than rare mutational changes. We have indirect evidence that expression of the habituated state, at least in pith cells, involves a metabolic feedback loop in which CDF induces its own production or inhibits its own degradation. When induced to undergo organogenesis, cells may change in their competence to respond to CDF by habituating. These changes are potentially reversible, appear to occur at high rates, involve shifts between two developmental states, and leave the cell totipotent. Changes in competence, therefore, are likely to have an epigenetic basis as well.

Our most interesting observation was that leaf tissues from plants of leaf and cortex origin differ in habituation. Moreover, the leaf phenotype characteristic of cortex plants was transmitted through the seed and segregated in a roughly 1:1 ratio. Although more detailed breeding experiments are needed to establish this, the results suggest that leaf- and cortex-derived plants differ genetically. Because these results were obtained with cells isolated from just one plant, however, it may be argued that, in this individual, the observed differences were due to a rare somatic mutation. To rule this out, we are now repeating these experiments with cells from several different plants. The important point is that our findings raise the possibility that stable developmental processes in plants may involve directed genetic changes.

Epigenetic changes are defined operationally and do not imply a specific cellular or molecular mechanism. The directed, reversible nature of these changes does not necessarily exclude a conservative genetic mechanism involving alterations in the sequential order, state, or number of genes in a chromosome. There is growing evidence for genetic changes of this type in eukaryotic organisms. Genetic studies with maize suggest that transposable genetic elements can migrate to new locations on the chromosome and regulate the transcription of nearby structural genes (McClintock, 1967; Nevers and Saedler, 1977). Transposable genes have also been proposed to account for mating-type determination in yeast (Hicks et al., 1977). Finally, there is direct, chemical evidence for rearrangement of DNA base sequences in antibody-producing cells of the mouse (Seidman and Leder, 1978) and for dramatic, heritable changes in the number of dihydrofolate reductase genes in cultured lines of mouse and Chinese hamster cells (Alt et al., 1978; Nunberg et al., 1978).

It is not known whether or not these mechanisms contribute to the observed variation of plant cells in culture. Cytokinin habituation provides an experimental system particularly well suited for studying this problem since there are hints that different states of habituation expressed by cells in culture are transmitted to the progeny of plants derived from these cells.

References

Alt, F. W.; Kellems, R. E.; Bertino, J. R.; and Schimke, R. T. (1978) Selective multiplication of dihydrofolate reductase genes in methotrexate-resistant variants of cultured murine cells. *J. Biol. Chem. 253;* 1357-70.

Beale, G. H. (1958) The role of the cytoplasm in antigen determination in *Paramecium aurelia. Proc. Roy. Soc. London B 148;* 308-14.

Binns, A., and Meins, F., Jr. (1973) Evidence that habituation of tobacco pith cells for cell division-promoting factors is heritable and potentially reversible. *Proc. Nat. Acad. Sci., U.S.A. 70;* 2660-62.

Binns, A. N., and F. Meins, Jr. (1979) Cold-sensitive expression of cytokinin habituation by tobacco pith cells in culture. *Planta 145,* 365-69.

Butcher, D. N. (1977) Secondary products in tissue culture. In *Plant Cell, Tissue and Organ Culture* (J. Reinert and Y. P. S. Bajaj, eds.), pp. 668-93. Springer-Verlag, Berlin, Heidelberg and New York.

Carlson, P. S. (1973) Methionine-sulfoximine-resistant mutants of tobacco. *Science 180;* 1366-68.

Carlson, P. S. (1974) Mitotic crossing-over in a higher plant. *Genet. Res. 24;* 109-12.

Cheng, T. Y. (1972) Induction of indole acetic acid synthetases in tobacco pith explants. *Plant Physiol. 50;* 723-27.

D'Amato, F. (1977) Cytogenetics of differentiation in tissue and cell cultures. In *Plant Cell, Tissue and Organ Culture* (J. Reinert and Y. P. S. Bajaj, eds.) pp. 344-57. Springer-Verlag, Berlin, Heidelberg, and New York.

Dyson, W. H., and Hall, R. H. (1972) N^6-(Δ^2-Isopentenyl) adenosine: Its occurrence as a free nucleoside in an autonomous strain of tobacco tissue. *Plant Physiol. 50;* 616-21.

Einset, J. W. (1977) Two effects of cytokinin on the auxin requirement of tobacco callus cultures. *Plant Physiol. 59;* 45-47.

Einset, J. W., and Skoog, F. (1973) Biosynthesis of cytokinins in cytokinin-autotrophic tobacco callus. *Proc. Nat. Acad. Sci., U.S.A. 70;* 658-60.

Feldman, L. J. (1975) Cytokinins and quiescent center activity in roots of *Zea.* In *Development and Function of Roots* (J. G. Torrey and D. T. Clarkson, eds.), pp. 55-72. Academic Press, New York.

Fox, J. E. (1963) Growth factor requirements and chromosome number in tobacco tissue cultures. *Physiol. Plantarum 16;* 793-803.

Gautheret, R. J. (1955) The nutrition of plant tissue cultures. *Ann. Rev. Plant Physiol. 6;* 433-84.

Gautheret, R. J. (1966) Factors affecting differentiation of plant tissues grown *in vitro.* In *Cell Differentiation and Morphogenesis,* pp. 55-96. North Holland Publishing Company, Amsterdam.

Gengenbach, B. G., Green, C. E., and Donovan, C. M. (1977) Inheritance of selected pathotoxin resistance in maize plants regenerated from cell cultures. *Proc. Nat. Acad. Sci., U.S.A. 74;* 5113-17.

Hicks, J. B.; Strathern, J. N.; and Herskowitz, I. (1977) The cassette model of mating-type interconversion. In *DNA Insertion Elements, Plasmids, and Episomes* (A. I. Bukhari, J. A. Shapiro, and S. L. Adhya, eds.), pp. 457-62. Cold Spring Harbor Laboratories, Cold Spring Harbor, New York.

Jablonski, J. R. and Skoog, F. (1954) Cell enlargement and cell division in excised tobacco pith tissue. *Physiol. Plantarum 7;* 16-24.

Kulescha, Z., and Gautheret, R. J. (1948) Sur l'élaboration de substances de croissance par 3 types de cultures de tissus de Scorsonere: Cultures normales, cultures de crown-gall et cultures accoutumees à l'heteroauxine. *Comp. Rend. Acad. Sci. Paris 227;* 292-94.

Lescure, A. M. (1973) Selection of markers of resistance to base-analogues in somatic cell cultures of *Nicotiana tabacum. Plant Sci. Lett. 1;* 375-83.

Linsmaier, E., and Skoog, F. (1965) Organic growth factor requirements of tobacco tissue cultures. *Physiol. Plantarum 18;* 100-127.

Luria, S. E., and Delbrück, M. (1943) Mutations of bacteria from virus sensitivity to virus resistance. *Genetics 28;* 491-511.

Lutz, A. (1971) Aptitudes morphogénétiques des cultures de tissus d'origine unicellulaire. In *Les Cultures de Tissus de Plantes. Colloq. Int. Cent. Nat. Rech. Scient.* Number 193, pp. 163-68. Éditions du Centre Nat. Rech. Scient., Paris.

McClintock, B. (1967) Genetic systems regulating gene expression during development. *Devel. Biol. Suppl. 1;* 84-112.

Maliga, P.; Sz.-Breznovits, A.; and Márton, L. (1973a) Streptomycin-resistant plants from callus culture of haploid tobacco. *Nature New Biol. 244;* 29-30.

Maliga, P.; Márton, L.; and Sz.-Breznovits, A. (1973b) 5-Bromodeoxyuridine-resistant cell lines from haploid tobacco. *Plant Sci. Lett. 1;* 119-21.

Meins, F., Jr. (1972) Stability of the tumor phenotype in crown-gall tumors of tobacco. *Prog. Exptl. Tumor Res. 15;* 93-109.

Meins, F., Jr. (1974) Mechanisms underlying the persistance of tumour autonomy in crown-gall disease. In *Tissue Culture and Plant Science 1974* (H. E. Street, ed.) pp. 233-64. Academic Press, London and New York.

Meins, F., Jr. (1977) Reversal of the neoplastic state in plants. *Am. J. Pathol. 89;* 687-702.

Meins, F., Jr., and Binns, A. N. (1977) Epigenetic variation of cultured somatic cells: Evidence for gradual changes in the requirement for factors promoting cell division. *Proc. Nat. Acad. Sci., U.S.A. 74;* 2928-32.

Meins, F., Jr., and Binns, A. N. (1978) Epigenetic clonal variation in the requirement of plant cells for cytokinins. In *The Clonal Basis of Development* (S. Subtelny and I. M. Sussex, eds.), pp. 185-201. Academic Press, New York.

Nanney, D. L. (1958) Epigenetic control systems. *Proc. Nat. Acad. Sci. U.S.A. 44;* 712-17.

Nevers, P., and Saedler, H. (1977) Transposable genetic elements as agents of gene instability and chromosomal rearrangements. *Nature 268;* 109-15.

Nunberg, J. H.; Kaufman, K. J.; Schimke, R. T.; Urlaub, G.; and Chasin, L. A. (1978) Amplified dihydrofolate reductase genes are localized to a homogeneously staining region of a single chromosome in a methotrexate-resistant Chinese hamster ovary cell line. *Proc. Nat. Acad. Sci., U.S.A. 75;* 5553-56.

Sand, S. A.; Sparrow, A. H.; and Smith, H. H. (1960) Chronic gamma irradiation effects on the mutable V and stable R loci of *Nicotiana. Genetics 45;* 289-308.

Seidman, J. G., and Leder, P. (1978) The arrangement and rearrangement of antibody genes. *Nature 276;* 790-95.

Skirvin, R. M. (1978) Natural and induced variation in tissue culture. *Euphytica 27;* 241-66.

Sung, Z. R. (1976) Mutagenesis of cultured plant cells. *Genetics 84;* 51-57.

Syōno, K., and Furuya, T. (1971) Effects of temperature on the cytokinin requirement of tobacco calluses. *Plant and Cell Physiol. 12;* 61-71.

Tandeau de Marsac. N., and Jouanneau, J.-P. (1972) Variation de l'exigence en cytokinine de lignées clonales de cellules de tabac. *Physiol. Vég. 10;* 369-80.

Thimann, K. V. (1969) The auxins. In *Physiology of Plant Growth and Development.* (M. B. Wilkins, ed.), pp. 3-45. McGraw-Hill, New York.

Widholm, J. M. (1974) Selection and characteristics of biochemical mutants of cultured plant cells. In *Tissue Culture and Plant Science 1974* (H. E. Street, ed.), pp. 287-99. Academic Press, New York.

Wood, H. N.; Braun, A. C.; Brandes, H.; and Kende, H. (1969) Studies on the distribution and properties of a new class of cell division promoting substances from higher plant species. *Proc. Nat. Acad. Sci., U.S.A. 62;* 349-56.

Index

Index